America's Voluntary Standards System
A 'Best Practice' Model for Asian Innovation Policies?

About the East-West Center

The East-West Center promotes better relations and understanding among the people and nations of the United States, Asia, and the Pacific through cooperative study, research, and dialogue. Established by the US Congress in 1960, the Center serves as a resource for information and analysis on critical issues of common concern, bringing people together to exchange views, build expertise, and develop policy options.

The Center's 21-acre Honolulu campus, adjacent to the University of Hawai'i at Mānoa, is located midway between Asia and the US mainland and features research, residential, and international conference facilities. The Center's Washington, DC, office focuses on preparing the United States for an era of growing Asia Pacific prominence.

The Center is an independent, public, nonprofit organization with funding from the US government, and additional support provided by private agencies, individuals, foundations, corporations, and governments in the region.

Policy Studies
an East-West Center series

Series Editors
Edward Aspinall and Dieter Ernst

Description
Policy Studies presents scholarly analysis of key contemporary domestic and international political, economic, and strategic issues affecting Asia in a policy relevant manner. Written for the policy community, academics, journalists, and the informed public, the peer-reviewed publications in this series provide new policy insights and perspectives based on extensive fieldwork and rigorous scholarship.

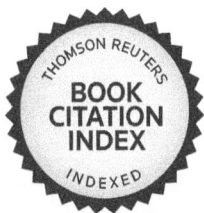

THOMSON REUTERS
BOOK
CITATION
INDEX
INDEXED

The East-West Center is pleased to announce that the Policy Studies series has been accepted for indexing in Web of Science Book Citation Index. The Web of Science is the largest and most comprehensive citation index available.

Notes to Contributors
Submissions may take the form of a proposal or complete manuscript. For more information on the Policy Studies series, please contact the Series Editors.

Editors, Policy Studies
East-West Center
1601 East-West Road
Honolulu, Hawai'i 96848-1601
Tel: 808.944.7197
Publications@EastWestCenter.org
EastWestCenter.org/PolicyStudies

Policy
Studies | 66

America's Voluntary Standards System

A 'Best Practice' Model for Asian Innovation Policies?

Dieter Ernst

America's Voluntary Standards System:
A 'Best Practice' Model for Asian Innovation Policies?
Dieter Ernst

ISSN 1547-1349 (print) and 1547-1330 (electronic)
ISBN 978-0-86638-239-7 (print) and 978-0-86638-240-3 (electronic)

The views expressed are those of the author(s) and not necessarily those of the Center.

Hard copies of all titles, and free electronic copies of most titles, are available from:

Publication Sales Office
East-West Center
1601 East-West Road
Honolulu, Hawai'i 96848-1601
Tel: 808.944.7145
Fax: 808.944.7376
EWCBooks@EastWestCenter.org
EastWestCenter.org/PolicyStudies

In Asia, hard copies of all titles, and electronic copies of select Southeast Asia titles, co-published in Singapore, are available from:

Institute of Southeast Asian Studies
30 Heng Mui Keng Terrace
Pasir Panjang Road, Singapore 119614
publish@iseas.edu.sg
bookshop.iseas.edu.sg

Contents

List of Acronyms

AIME   American Institute of Mining, Metallurgical, and Petroleum Engineers, Inc.

ANSI   American National Standards Institute

ARPANET   Advanced Research Projects Agency Network [of the US Department of Defense—the world's first operational packet-switching network and the forerunner of the Internet]

ARRA   American Recovery and Reinvestment Act [of 2009, US]

ASCE   American Society of Civil Engineers

ASME   [ASME, formerly the] American Society of Mechanical Engineers

ASTM   [ASTM International, formerly the] American Society for Testing and Materials

AT&T   [AT&T Corporation, formerly the] American Telephone and Telegraph Corporation

BSI   The British Standards Institution

DIN   Deutsches Institut für Normung e.V. [the German Institute for Standardization]

DOE	Department of Energy [US]
DOJ	Department of Justice [US]
EISA	Energy Independence and Security Act [of 2007, US]
EU	European Union
FERC	Federal Energy Regulatory Commission
FRAND	fair, reasonable, and non-discriminatory [patent licensing conditions]
GE	General Electric
IAB	Internet Architecture Board
IBM	International Business Machines
ICT	information and communications technology
IEC	International Electrotechnical Commission [of the International Organization for Standardization]
IEEE	Institute of Electrical and Electronics Engineers
IESG	Internet Engineering Steering Group
IETF	Internet Engineering Task Force
INCITS	International Committee for Information Technology Standards
IP	Internet protocol
IPv6	Internet protocol version 6
ISO	International Organization for Standardization
IT	information technology
ITIC	Information Technology Industry Council
JTC1	Joint Technical Committee 1 [of the ISO and the IEC]

NEMA	National Electrical Manufacturers Association
NERC	North American Electric Reliability Corporation
NIST	National Institute of Standards and Technology [of the US Department of Commerce]
NTTAA	National Technology Transfer and Advancement Act [of 1995, US]
OASIS	Organization for the Advancement of Structured Information Standards [Consortium; of which IBM, Oracle, and Sun Microsystems are strong supporters]
ODF	Open Document Format for Office Applications [by OASIS]
OEDER	Office of Electricity Delivery and Energy Reliability [of DOE]
OMB	Office of Management and Budget [US]
OOXML	Office Open Extensible Markup Language [by Microsoft]
OSTP	Office of Science and Technology Policy [of the Executive Office of the President, US]
OTA	Office of Technology Assessment [of the Congress, US]
R&D	research and development
SAC	Standardization Administration of China
SDOs	standards-developing organizations [international-industry-based, e.g., the IEEE]
SFF-SIG	Small Form Factor Special Interest Group
SGIP	Smart Grid Interoperability Panel
SSOs	standards-setting organizations [private, e.g., business consortia]

TCP/IP	transmission-control protocol/Internet protocol [communications protocol used for the Internet and other similar networks—covering both the TCP and the IP, as the first two networking protocols defined in this standard, and now including multiple additional protocols for handling data communication]
USTR	United States Trade Representative [Office of]
VITA	[VITA, formerly the] VMEbus International Trade Association [open-standards-development trade organization]
VSO	VITA's Standards Organization [ANSI-accredited organization]
W3C	World Wide Web Consortium
WTO	World Trade Organization
ZigBee	a specification for a suite of high-level communication protocols using small, low-power digital radios based on an IEEE 802 standard for personal area networks

Executive Summary

Across Asia there is a keen interest in the potential advantages of America's market-led system of voluntary standards and its contribution to US innovation leadership in complex technologies.

While Asian interest is strong, there is a recognized lack of information concerning America's voluntary standards system and the commercial, historical, political, and sociological realities of how it began and now functions. Who controls the objectives, who provides the resources, how are decisions actually reached, what feedback and appeal structures exist, and what is the true role of the US government within the American standards system? What are the system's comparative strengths and weaknesses—especially when contrasted with government-led alternatives, such as in China?

For its proponents, America's voluntary standards system is an effective response to new challenges faced by innovation policy in the global knowledge economy. Hence, this system may serve as a "best practice" model for other countries. In this view the key to success is a bottom-up, decentralized, informal, market-led approach that provides "open" access and responds quickly to the ever-accelerating pace of technical change and the sometimes disruptive shifts in markets.

The view from outside the United States is different. Interviews with standardization

> *For US proponents, the key to success is a bottom-up, decentralized, informal, market-led approach to standardization*

experts from China, Europe, Korea, and Taiwan highlight serious concerns about the possible draw-backs of a standards system that is largely driven by the private sector. There are doubts, especially in China, about whether the American system can properly balance public and private interests in times of extraordinary national and global challenges.

To assess these conflicting perceptions this study examines the defining characteristics of the American standards system. In addition to a review of the relevant literature, the study draws on more than 70 interviews conducted since 2009 with standards developers, implementers, and users in the United States as well as discussions (by Internet and phone) with leading standardization experts. The analysis highlights the deeply rooted US tradition of bottom-up, decentralized, informal, market-led, governance of standards development and clarifies the often-neglected role of the US government in identifying these standards.

First, this study considers how history has shaped and defined the unique trajectory of the American system. This consensus-driven voluntary standards system is grounded in a tradition of local self-government. A unique mix of individualism, local control, meritocracy, and voluntarism gave rise to a deeply entrenched preference for the private coordination of economic activity. This offers a partial explanation of why the United States never established a centralized authority responsible for creating and enforcing standards.

Second, this study considers how America's decentralized governance of standards development, while messy, in principle gives voice to a diversity of opinions and approaches and hence provides benefits unavailable to top-down, command-style, government-centered standards systems. The study provides two illustrative examples of such benefits: the Internet Engineering Task Force (IETF) model of system-level standards development for the Internet and the outsourcing of detailed component specification to informal peer-group networks in the information technology (IT) industry.

Third, despite the extraordinary strengths of the market-led approach to standardization, the study highlights important drawbacks of the American system. America's voluntary standards system is prone to intense conflicts; its capacity to coordinate the strategies of diverse standardization stakeholders is limited; it lacks sufficient openness and transparency in standards development; and it fails to provide equal access to standards development for all stakeholders including small- and medium-sized enterprises and users. Illustrative examples document the use of "essential patents" as strategic weapons to delay, obstruct, or prohibit standardization processes.

Fourth, this study documents the role of the US government—often a "missing link" in discussions of the American standards system—and highlights both successes and failures. In contrast to widespread perceptions, the US government has frequently been an active participant in standards development, primarily through antitrust policies and public procurement. After a period of passivity, the government now appears to be returning to a more activist approach—attempting to facilitate and strengthen public-private standards-development partnerships.

A case study of the Smart Grid interoperability standards project, coordinated by the National Institute of Standards and Technology (NIST), provides an important example. This project is distinguished by the prominent role of government agencies in shaping the agenda and in providing key resources and controlling project outcomes.

Globalization and the increasing complexity of advanced technology imply that the role of US public policy needs to be strengthened. Especially for contested issues like standards-essential patents, the government needs to act as an enabler, coordinator and, if necessary, an enforcer of the rules of the game to prevent the abuse of market power.

Fifth, this study proposes answers of direct relevance to Asian countries. America's system of standardization reflects the unique characteristics of its history and economic institutions. Thus, even if the strengths outweigh the weaknesses of the American standards system—which this study argues is the case—it would still be hard to fully replicate that system in other countries. This is especially true for Asian countries with their different economic histories and institutions. Of great interest for policymakers in Asian countries,

> *Of interest for Asian policymakers will be current attempts to establish and strengthen robust public-private standards-development partnerships*

and especially for those in China, will be the current attempts in the United States to establish and strengthen robust public-private standards-development partnerships. As illustrated by the US Smart Grid interoperability standards project, public-private partnerships may indeed provide a "best practice" model enabling both consumers and communities to play a more active role in formulating standards.

In conclusion this study provides an important message for standards and innovation policies in Asia. Attempts to copy and replicate the American standards system will face clear limitations. While standards systems everywhere are confronted with similar tasks, there are significant differences in the organization and governance of standardization processes. These differences reflect the unique characteristics of each country's differing economic institutions, their levels of development, their economic-growth models, and their cultures and history.

America's Voluntary Standards System

A 'Best Practice' Model for Asian Innovation Policies?[1]

Introduction

Across Asia there is a keen interest in the potential advantages of America's market-led system of voluntary standards and its contribution to the country's superior innovation capacity. The US approach to standardization has enabled US innovation leadership in complex technology networks like the Internet and the World Wide Web. This approach has provided right-on-time flexible interoperability standards that make it possible to combine a variety of components into functional systems.

But little is known in Asia about how America's standards system really works in practice. When earlier versions of this study were presented in various Asian countries there were numerous questions concerning the objectives, the governance ("Who controls strategic resources and shapes decisions on the US standardization strategy?"), the role of government, and the unique strengths and weaknesses of the US market-led system of voluntary standards. This study seeks to provide answers to many of these questions.

The US standards system is focused on voluntary consensus standards that are created by private-sector standards-development organizations. For its proponents, the US system is an effective response to the new challenges that innovation policy faces in the global knowledge economy and, hence, can serve as a "best practice" model for other countries. In this view the key to success is a bottom-up, decentralized, informal, market-led approach providing "open" access and responding quickly to the rapid pace of technical change and the sometimes disruptive shifts in markets.

The view from outside the United States is different. Interviews with standardization experts from China, Korea, Taiwan, and Europe show that these foreign observers are well aware of the extraordinary achievements of the US market-led system of voluntary standards in generating innovation.[2] There is a keen interest to learn more about the potential advantages of a US-style voluntary standards system and how this system really works in practice.

But these foreign observers also express serious concerns about possible drawbacks of a standards system that is largely driven by the private sector. There are doubts, especially in China, whether the US system can balance public and private interests in times of extraordinary national and global challenges. In attempting to assess the merits of these conflicting perceptions this study examines defining characteristics of the American standards system. Beyond reviewing the relevant literature, this study draws on more than 70 interviews conducted since 2009 in the United States with standards developers, implementers, and users, as well as discussions (by Internet and phone) with leading US and international standardization experts.

This study offers two basic propositions:

First, that the deeply rooted US tradition of bottom-up, decentralized, informal, market-led standardization has been extraordinarily successful in generating innovation (National Science Board 2012; Institute for Defense Analyses 2012). The predominance of the private sector has clearly fostered entrepreneurship and risk-taking. However, after the recent global economic crisis, new questions have been raised concerning whether the incentives for (sometimes excessive) risk-taking built into the US market-led standards system need to be countered by forces (including stricter regulations) emphasizing careful assessment of the broader risks and social costs of innovation.

It may not be at all easy to transplant the US standards system to other countries

Second, that America's system of standardization is a microcosm of US-style capitalism. As the literature on the varieties of capitalism convincingly demonstrates, convergence among different varieties of capitalism is limited (Hall and Soskice 2001)—partial convergence often goes hand-in-hand with persistent diversity (Ernst and Raven-hill 2000). On this basis it may not be at all easy to transplant the US standards system to other countries. The decentralized voluntary US standards system is deeply embedded in "American political culture and the manner in which industrialization took place in the United States" (Office of Technology Assessment 1992, 39).

The private sector has long been developing de facto (existing, but not necessarily legally ordained) voluntary consensus standards, either within individual firms or through inter-firm standardizations using both formal and informal processes.

However the US government has also played an important role, both behind the scenes and directly, in shaping the evolution and the defining characteristics of the American standards system. This has given rise to a unique form of public-private interaction which, for many foreign observers, is the less-well-known aspect of the US standards system. Of particular interest is that the boundaries set for the role of the government in standardization have moved over time, in line with shifts away from the welfare and warfare state to the deregulation of markets.

To examine the defining characteristics of the current US standards system and its strengths and weaknesses, this study will proceed as follows:

"Expectations" reviews widely shared expectations in the United States that the US system can serve as a "best practice model" for fostering innovation—and that this model can be replicated in other countries.

"Defining Characteristics of the American Standards System" examines the US voluntary consensus standards system that is grounded in a tradition of local self-government, highlighting the historical roots of these systems and their unique strengths. To illustrate the benefits of the voluntary standards system, two examples are examined—the Internet Engineering Task Force (IETF) model of system-level standards development for the Internet and the outsourcing of detailed component specifications to informal peer-group networks in the information technology (IT) industry.

"What Are the Drawbacks of the American System?" reviews the literature addressing the drawbacks of the US voluntary standards system. The analysis explores why America's voluntary standards system is prone to intense conflicts and why it lacks the capacity to coordinate diverse standardization stakeholders—a capacity needed for an integrated national innovation policy. "The Role of the US Government" examines the missing link of the US standards system—i.e., the important role of the US government—and documents the successes and failures of the government's efforts to establish robust public-private standards-development partnerships.

"A New Approach to Public-Private Standardization Partnerships? The Smart Grid Interoperability Standards Project" analyzes this project—coordinated by the National Institute of Standards and Technology (NIST)—as an illustrative example of such partnerships.

The study concludes with a summary of main findings and highlights policy implications with a focus on possible lessons for Asian innovation policies.

Expectations

Proponents of the US system believe that a "voluntary standards system" is capable of accomplishing innovation-policy objectives better than any other standards system, especially systems that are heavily reliant on the government. The bottom-up, decentralized, informal, market-led US approach to standardization has indeed been extraordinarily successful in generating innovation in products as well as in processes, services, and software, especially in the IT industry. The US system has also enabled companies to respond in time to the accelerating pace of innovation.

As a result, it is argued, the US standards system should serve as a "best practice model" and other countries should strive to replicate the voluntary standards system.

It is worth noting the response of the American National Standards Institute (ANSI) to a national survey on the impact of globalization on US standards policies. According to ANSI, "No change to the current private sector-led and public sector-supported standardization system is warranted…as…the current system works *well* [emphasis in original]" (ANSI 2009, 1). And the US Commerce Department's International Trade Administration argues that "the voluntary standards system has been a key to driving technical innovation and maintaining the United States' position as a global leader in technology…in today's global economy" (International Trade Administration 2009, 2).

For the proponents, the main asset of the American standards system is its bottom-up, decentralized, informal, market-led approach that provides "open" access. In a recent paper, Chuck Powers—a Motorola engineer who is a highly respected participant in US and international standards bodies—defines "open" access as follows: "Anyone can participate, can work to achieve results, can bring perspectives, and can work to achieve consensus. And it is not just big [intellectual property] holders, as there are also a lot of small companies represented, individuals, universities, etc." (Powers 2009, 9).

Proponents of the US voluntary standards system believe that it accomplishes innovation-policy objectives better than any other standards system

This study's interviews show that many observers in Asia find it difficult to accept such optimistic claims. Some interviewees acknowledge that this concept of open and equal access may well exist within informal peer-group networks of dedicated engineers whose overriding interest is to create something new and to get this job done as quickly as possible and without much fuss. Most Asian observers, however, remain skeptical that really open access can be realized in industries that are shaped by intensive technology-based competition.

Competition in the IT industry is, in fact, shaped by brutal rivalries and battles among leading players (Ernst 2002). Success or failure in the rapidly moving IT industry is defined by return-on-investment and speed-to-market and every business function, including research and development (R&D) and standard development, is measured by these criteria. Under such conditions the proposition of equal access to standards-developing organizations (SDOs) may be more wish than reality.[3] As observed by the *Economist*, "In the computer industry, new standards can be the source of enormous wealth, or the death of corporate empires. With so much at stake, standards arouse violent passions. Much of the propaganda pumped out by individual firms is aimed at convincing customers and other firms that their product has become a 'standard.'"[4] Companies have very little room for compromise on sharing the potentially significant economic rents to be reaped by those who shape and control the process of standardization.

A second broadly shared expectation in the United States is that the American standards system should serve as a "best practice model" and that it can be replicated in other countries. What constitutes best practice can be determined through a process of benchmarking. For standardization, the "best practice model" would imply that a standards-development organization in another country is supposed to benefit from a process of progressive compliance ("convergence") with key elements of the US voluntary standards system.

> *New standards can be the source of enormous wealth, or the death of corporate empires. With so much at stake, standards arouse violent passions*

This expectation can be found, for instance, in ANSI's *United States Standards Strategy*, last revised in 2010 (ANSI 2010c). This document proposes the "universal application of the globally accepted principles for development of global standards" based on the US voluntary standards system. The document states:

> Open and accessible, the US standardization system has contributed its technology, in gigantic proportions, to other standardization models and to other societies. It is committed, not only to interests within its own territory, but to international standardization, and to a global trading system that is balanced and without obstacles. This strategy is designed to strengthen the standards system of the United States and all who benefit from it. (ANSI 2010c, vi)

A similar optimism was expressed by Chuck Powers:

> The US standards system is healthy and robust because there is a careful balance of competing interest; the open process ensures the system is appealing to all stakeholders, without unnecessary requirements; the US government plays a limited role and simply ensures a level playing field; and there are numerous examples

of successful US standards deployed around the globe. It is also scalable because it can balance competing interests globally, not just in the [United States]. (Powers 2009, 10)

But is it realistic to assume that, over time, other countries, including China, will converge on the US-style market-led standardization system as the "best practice" model?

China's primary concern is to develop this vast quasi-continental country as rapidly as possible and to achieve the productivity and income levels of the European Union (EU), Japan, and the United States (Ernst 2011). Strengthening China's domestic innovative capacity is considered to be the key to a sustainable transformation of its economy beyond the export-oriented "global factory" model. To achieve this goal, China's government is very serious in its aspiration to move from being a mere *standard-taker* to become a *co-shaper*, and, in some areas, a *lead shaper* of international standards.

From the Chinese perspective, reducing dependence on manufactured exports will only be possible if China succeeds in strengthening its domestic innovative capacity. To achieve this objective China seeks to upgrade its standards system to lessen the "control of foreign advanced countries over the [People's Republic of China]," especially "in the area of high and new technology," and increase the effectiveness of Chinese technical standards as important protective measures or barriers to "relieve the adverse impact of foreign products on the China market" (SAC 2004; preface and part I, section IV). This document by the Standardization Administration of China (SAC) adds that China's standardization strategy needs to fill a policy vacuum, as China's accession commitments to the World Trade Organization (WTO) have substantially reduced the use of most other trade restrictions such as import quotas, licensing requirements, and tariffs.[5]

In short, there are vastly different perceptions in the United States and in China as to what constitutes legitimate goals of innovation and of standards policies.

While this study is focused on the US standards system, its findings support an important proposition for future comparative research: While standards everywhere are confronted with similar tasks, there are significant differences in the organization and governance of standardization processes. These differences reflect the unique characteristics of each country's differing economic institutions, their levels of development, their economic-growth models, and their cultures and history (Kindleberger 1983, 383).

Defining Characteristics of the American Standards System

Evolving Tasks of Standardization

There are an almost infinite number of standards that differ in their form and purpose. To shed light on the evolving tasks of standardization this study will examine standards as a concept and introduce an operational definition. A state-of-the-art definition of technical standards is provided by NIST as part of its Smart Grid interoperability standards project (NIST 2010, 19–20): Standards are

> specifications that establish the fitness of a product for a particular use or that define the function and performance of a device or system. Standards are key facilitators of compatibility and interoperability....Interoperability...[is]...the capability of two or more networks, systems, devices, applications, or components to exchange and readily use...meaningful, actionable information—securely, effectively, and with little or no inconvenience to the user....[Standards] define specifications for languages, communication protocols, data formats, linkages within and across systems, interfaces between software applications and between hardware devices, and much more. Standards must be robust so that they can be extended to accommodate future applications and technologies.

In the literature, standards are normally categorized as "proprietary" versus "open" and as "de facto" versus "de jure" (Stango 2004). Proprietary standards are owned by a company which may license them to others while open standards "are available to all potential users, usually without fee" (Steinfield et al. 2007, 163). De facto standards achieve adoption through standards competition among rival standards consortia. Finally, de jure standards are adopted through consensus, which is sometimes formally expressed through industry committees or formal standards organizations.

At the most fundamental level, standards help to ensure the quality and safety of production processes, products, and services and to prevent negative impacts on health and the environment. An important function of standards is to reduce "risks for makers of compliant products and users of these products" (Alderman 2009, 2–3).

Standards are necessary to reap the growth and productivity benefits of increasing specialization. This was historically analyzed in the chapter entitled "That the Division of Labor is Limited by the Extent of the Market" of Adam Smith's *An Inquiry Into the Nature and Causes of the Wealth of Nations* (Smith 1776, Book 1, chapter III). According to economic historian Charles Kindleberger, "For the most part, standardization was originally undertaken by merchants" to facilitate a progressive specialization through trade (Kindleberger 1983, 378–9).

Today, however, specialization extends well beyond trade into manufacturing and services. This includes engineering, product development, and research. Equally important is the international dimension. As globalization has extended beyond markets for goods and finance into markets for technology and knowledge workers, standards are no longer restricted to national boundaries.

Standards have become a critical enabler of international trade and investment—they facilitate data exchange as well as knowledge sharing among geographically dispersed participants within global corporate networks of innovation and production (Ernst 2005b; Ernst 2005c). As network sociologists emphasize, the "creation and diffusion of standards underlying new technologies is a driving element of contemporary globalization" (Grewal 2008, 194).

In short, standards are the lifeblood of innovation in the global knowledge economy. Today standards are necessary not only to reap economies of scale and scope but also to reduce transaction costs and to minimize possible duplication of efforts. Standards are required to enable data transfer and knowledge exchange and to facilitate interoperability of components and software within increasingly complex technology systems (e.g., smart phones and switching systems).

Without interoperability standards it would be impossible to achieve "network externalities" which shape competition

Standards are the lifeblood of innovation in the global knowledge economy

in markets for products and services using information and communication technologies (Katz and Shapiro 1985). In these markets, "as the set of users expands, each user benefits from being able to communicate with more persons (who have become users of the product or service)" (Rohlfs 2001, 8). "Network externalities" imply that companies succeed "when customers expect that the installed base of...[the company's]...technology [will] become larger than any other" with the result that customers "adopt that technology to the virtual exclusion of others" (Sheremata 2004, 359).

To cope with these critical challenges, standardization has become a complex and multi-layered activity involving multiple stakeholders with different capabilities, objectives, resources, and strategies. In the United States stakeholders are primarily from the private for-profit sector but also include government agencies and non-profit organizations such as universities, research labs, and non-governmental organizations.

Importantly, standardization is a knowledge-intensive activity requiring well-educated and experienced engineers and other professionals. While engineers originally created this discipline, key concepts are now shaped by legal counselors as well as by corporate executives and government officials. Equally important are the considerable financial resources required to develop and implement effective standards.

Significant differences exist in the governance and organization of standardization processes. As noted earlier, these differences reflect the individual characteristics of different countries' economic institutions, their levels of development, their economic-growth models and their cultures and history. An unfortunate weakness of the current literature on standardization is that it lacks systematic comparisons of the multiple different national standards systems and their divergent development trajectories.[6] Existing comparative studies have focused on comparisons of the European and the US systems, neglecting important developments in Brazil, India, Japan (Yamauchi 2004),[7] Russia, and, most importantly, China.[8]

Differences in standardization processes also reflect the diversity in the underlying conditions of population, products, resources, and tastes. Finally, and most importantly, standardization processes differ across industrial sectors reflecting differences in competitive dynamics, demand patterns, and technology.

In theory it would be desirable if all formal standards-development bodies and consortia would: assure fair, reasonable, and non-discriminatory (FRAND) patent-licensing conditions; enforce early disclosure of essential patents; foster the unhindered application of standards; and prevent the blocking of standards.[9] In reality, as demonstrated below, there are many conflicting interests and diverse strategies and organizational approaches.

Historical Roots

To better understand the strengths and weaknesses of the current US standards system there is a need to go back to the early nineteenth century, when the nation entered the industrial stage.[10] In contrast to many other countries, where unified national standards bodies were established, "standards development organizations in the United States first emerged in the private sector, in response to specific needs and concerns" (Office of Technology Assessment 1992, 39).

Early pioneers in US standardization were scientific and technical societies (e.g., the American Society of Civil Engineers [ASCE], established in 1852; the American Society for Mechanical Engineers [ASME], established in 1880; and the American Society for Testing and Materials [ASTM; now ASTM International], established in 1898) and trade associations (such as the American Iron and Steel Institute, established in 1855). From their very beginnings these societies and associations established their right to make their own standards.

The American Society for Mechanical Engineers has a tradition of generating publications tracing the evolution of their society (Ferguson 1974, Sinclair 1980). ASME was founded by prominent engineers in 1880, a time when US engineering schools and institutions were rapidly expanding. As engineering was then still a relatively new profession, "engineers of the day moved easily among the concerns of civil, industrial, mechanical, and mining engineering" (ASME 2010, 1).

Steam power was, at that time, the dominant technology, driving locomotives and ships and factory and mine equipment and machinery. When boiler explosions began to multiply, the spectacular accidents aroused public outcries for improving the safety of boilers and related equipment. A Boiler Code Committee was formed in 1911 that led to the Boiler Code being published in 1914–15 and later incorporated into laws of most US states and territories and Canadian provinces.

These and other early US standards (e.g., standards for building codes and fire equipment) "were driven by public pressure and the ethical concerns of the engineering profession. Standardization was a solution demanded by public concern and professional responsibility" (Spring 2009, 6). Interviews with today's US engineers involved in standardization indicated there is still a very strong sense of these original motivations. Arguably this provides one rationale for why US engineers remain so deeply attached to the US voluntary standards system with its long tradition of decentralized decision making.

Chuck Powers describes the US standards system as "a highly successful system because it is driven from the bottom-up" (Powers 2009, 9). This belief is often associated with a deep distrust of government-centered standards systems.

A major catalyst for the emerging US standards system was the significant standardization effort required to interconnect America's railways. Economic historian Alfred D. Chandler notes that cooperation between business enterprises "was essential for the creation of an integrated national transportation network. Without such cooperation the standardization of equipment and operating procedures required to move through passengers and freight quickly and efficiently from one line to another would have been much slower in coming" (Chandler 1977, 143). By 1897, 1,158 independent railroad companies had laid and interconnected over 240,000 miles of track with minimal technical assistance from the US government. This required not only the industry-wide standardization of track gauges but also of cars and their equipment, uniform procedures and freight classifications, and standardized time references.

A major catalyst for the emerging US standards system was the significant standardization effort required to interconnect America's railways

This achievement left a powerful legacy for US economic philosophy—it helps explain why, until today, the defining characteristic of the US standardization system is "a strong political and cultural bias in favor of the marketplace" (Office of Technology Assessment 1992, 39).[11]

As Carl Cargill puts it in his important 1989 study, the US standardization system is built on voluntary standards, developed by engineers, "to make the industry grow or to make it profitable and/or less complex" (Cargill 1989, 21). In this view the role of government is to provide a limited set of regulations to guarantee the safety and welfare of its citizens and to prevent the abuse of market power.

Decentralized Self-Government

Another defining characteristic of the US standards system is the recognition that it has been shaped by the fundamental political traditions of the American Revolution. A unique mix of individualism, local control, meritocracy, and voluntarism gave rise to a deeply entrenched preference for the private coordination of economic activity (Garcia 1992).

An important institutional innovation dates back to 1916. By then the proliferation of engineering societies had led to considerable confusion among users of standards on acceptability and concerns about inconsistent quality. To respond to these problems of uncoordinated competition among engineering societies the American Institute of Electrical Engineers (now the Institute of Electrical and Electronics Engineers [IEEE]) invited the American Society of Mechanical Engineers (ASME), the American Society of Civil Engineers (ASCE), the American Institute of Mining and Metallurgical Engineers (AIME) and the American Society for Testing Materials (now ASTM International) to join in establishing an impartial national body to coordinate standards development, approve national consensus standards, and reduce user confusion regarding acceptability. These five private organizations subsequently invited the US Departments of Commerce, Navy, and War to join them as founders (ANSI 2010b).

To transform industry standards into national standards, the American Society of Civil Engineers (ASCE) developed a federation of "industrial legislatures" intended to manifest a political philosophy in support of the directness and vitality of elementary local self-government (Russell 2006, 74–76). That philosophy is nicely captured in an article by the ASCE's first full-time secretary, Paul Agnew:

> We do not leave to Congress...the decision whether a bridge shall be built in the city of Oshkosh. We leave it to the people

of Oshkosh, who will walk over it and ride over it, and who will have to pay for it. Why should not the very limited groups directly interested in each of the innumerable industrial problems with which they are faced, themselves solve these problems through cooperative effort? (Agnew 1926, 95)

Resistance to Regulatory Standards

The fundamental US orientation towards decentralized self-government explains why—in contrast to countries like France, Germany, Japan, and, now, China—the United States "has never established a centralized, overarching authority responsible for creating and enforcing standards" (Russell 2006, 77).

There is a widespread consensus in the United States that government regulatory standards are a "poor substitute" for voluntary, market-driven, standards, and that government regulatory standards are apt to stifle entrepreneurship and innovation. To quote again Carl Cargill, government regulatory standards "are ponderous, like a juggernaut, they are hard to start and steer, require vast throngs of people to keep them moving, and seem to acquire a life of their own once they get going—once rolling, they are usually difficult to stop" (Cargill 1989, 18).

The US resistance to more active government involvement through regulatory standards is deeply entrenched. Leading industry representatives testifying at a 1990 NIST hearing on the role of the federal government in standardization were emphatic in their resistance to a more active governmental role (Mattli and Buethe 2003, 24). More recently this reluctance to accept a broader role for the federal government is stated again in the 2012 Office of Science and Technology Policy (OSTP; of the Executive Office of the President) letter to US agencies outlining federal policy regarding the involvement of private industry in standard-setting.[12]

And a study, entitled *Risky Business: The US Software Industry's Perspective on US Government Engagement in the Process of Standard Setting* finds that key players in that industry believe that the existing governance mechanisms for standards development are adequate; doubt the US software industry can agree on a consensus strategy regarding the proper role of US government (because of conflicting strategic interests); and are not interested in developing a more structured approach to the governance of standards development (Lord 2007).[13]

For non-US observers the resultant institutional heterogeneity and fragmentation may look like chaos. But for Americans the principles of consensus and pluralistic governance through local self-government are deeply familiar concepts and are part of their cultural heritage.

Advantages of Decentralized Self-Government

The potential advantages of decentralized self-government are well-established in theories of innovation and organization.

Complexity theory is now an integral part of innovation theory. For complexity theorists, decentralized and flexible institutions, developed by participants who are "intimately knowledgeable about details of their activities, are likely to be more workable than blueprints developed by policy analysts and imposed by politicians and bureaucrats" (Axelrod and Coehen 1999, 22).

Contemporary innovation theory emphasizes that innovation results from interactions of multiple and diverse stakeholders through geographically dispersed innovation networks. Thus innovation requires "complex systems that are characterized by the heterogeneity of agents with different functions, different endowments, different learning capabilities and different perspectives, and most important different locations in the multidimensional spaces of geography, knowledge, technology, and reputation" (Antonelli 2011).

> *Decentralized and flexible institutions are more workable than blueprints developed by policy analysts and imposed by politicians and bureaucrats*

The vision of local self-government finds ample support in the "collective action" governance theory developed by Elinor Ostrom, the 2009 Nobel laureate in economics. In her path-breaking study *Governing the Commons: The Evolution of Institutions for Collective Action,* Ostrom argues that "all organizational arrangements are subject to stress, weakness, and failure" (Ostrom 1990, 25). However external regulatory agencies are even more subject to stress, weakness, and failure: "A regulatory agency...always needs to hire its own monitors. The regulatory agency then faces the principal-agent problem of how to ensure that the monitors do their own job....It is difficult for a central authority to have sufficient time-and-place information to estimate accurately both the carrying capacity of a...[public good, like standards]...and the appropriate ...[incentives and fines]...to induce cooperative behavior" (Ostrom 1990, 17).

Example 1: The IETF Model of System-Level Standardization for the Internet

The history of the Internet provides important insights into the potential strengths of the American system of decentralized governance of technology development and standardization. It also highlights the sometimes messy and often unpredictable evolution of public-private interaction.

Janet Abbate, in her path-breaking study *Inventing the Internet,* examines the forces that transformed the Advanced Research Projects Agency Network (ARPANET), as it was initially implemented under the auspices of the US Department of Defense, into the heterogeneous and decentralized "network of networks" known today as the Internet. While initially "The Internet...reflected the command economy of military procurement,...the key to the Internet's later commercial success was that the project internalized the competitive forces of the market by bringing representatives of diverse interest groups together and allowing them to argue through design issues" (Abbate 1999, 145).

Key elements of the Internet's decentralized self-governance were reflected in a commitment to flexibility and diversity—not only in the technical design of the Internet's architecture but also in its implementation and in the process of developing the fundamental standards. The IETF was an important battleground for many decisions on balancing flexibility and diversity with the initial philosophy of "mission-oriented research" that had shaped the ARPANET. The history of IETF, and its unique approach to the strategy and organization of standards development, provides us with a micro-view of the potential strengths of a decentralized model of self-governance.

While IETF is an international standards organization, from its beginnings it has been imbued with the values of the US Internet pioneers, i.e., a basic presumption that diversity of opinions and approaches is preferable to top-down, command-style, "mission-oriented research" governance. IETF develops and promotes Internet standards,[14] cooperating closely with the World Wide Web Consortium (W3C) and International Organization for Standardization (ISO)/International Electrotechnical Commission (IEC) standards bodies[15] and dealing in particular with standards of the transmission-control protocol/Internet protocol (TCP/IP) suite.[16] IETF defines itself as an open-standards organization with no formal membership or membership requirements. Though their work is usually funded by their employers or sponsors, all participants and managers are volunteers.

IETF is organized into numerous working groups and informal discussion groups, each dealing with a specific topic. Each group is intended to complete work on its selected topic and then disband. Each working group has a charter describing its focus and what it is expected to produce and when it should complete its task and appoints a chairperson or multiple co-chairs. Working groups are organized into areas such as applications; Internet operations and management; real-time applications and infrastructure; and routing, security, and transport. Area directors, together with the IETF chair, form the Internet Engineering Steering Group (IESG) responsible for the overall operation of the IETF.

In principle the process of creating an Internet standard is straightforward. A specification undergoes a period of development and several iterations of review by the Internet community and revision based upon experience. It is then adopted as a draft standard by the IESG, and published.

In practice, however, the process is much more complicated. As is described in *The Internet Standards Process: Best Current Practice*, this is "due to (1) the difficulty of creating specifications of high technical quality; (2) the need to consider the interests of all of the affected parties; (3) the importance of establishing widespread community consensus; and (4) the difficulty of evaluating the utility of a particular specification for the Internet community" (Bradner 1996). Hoffman (2009) comes to a similar conclusion.

IETF's insistence on openness and flexibility has significant trade-offs. One is the need to devise detailed and cumbersome procedures for conflict resolution and appeals. This is an indication of the fundamental dilemma inherent in the model of largely self-governed standards development. While significant time and effort is required for implementation and testing and to allow all interested parties to comment, today's rapid development of networking technology demands an equally rapid development of standards. Over time this conflict has become increasingly serious—especially with the current challenge to transition to a new generation of Internet architecture (e.g., Internet protocol version 6 [IPv6], responding to the impending scarcity of available IPv4 Internet addresses).

Thus far, attempts within IETF to speed up the standardization process have produced mixed results (Simcoe 2007, DeNardis 2009). An additional concern is that, over the last few years, corporate interests have gained considerably in their influence. As observed by Abbate in her study of the transition from the ARPANET to the Internet, after the late 1970s "the Internet and its creators were no longer operating in the insulated world of defense research; they had entered the arena of commerce and international politics, and supporters of the Internet technology would have to adapt to this new reality" (Abbate 1999, 153).

Since Abbate's 1999 publication, the influence of leading corporations has only increased. IETF meetings attract more and more participants, substantially increasing the cost of running the meetings. As a result IETF must increasingly rely on corporate sponsorships.

Finally and most importantly, IETF faces increasing difficulties in attempting to adjust its policies on intellectual property rights to increasingly technology-centered global competition and the ever-more-aggressive corporate   tactics of "strategic patenting" by leading IT corporations. While the IETF model of decentralized self-governance was initially an important organizational innovation it may now face increasing limitations reflecting the cutthroat competition in this critical sector of the IT industry.

> *While the IETF model was an important innovation, it may now face increasing limitations reflecting the cutthroat competition in this critical sector*

Example 2: Outsourcing of Component Specification

Outsourcing of detailed component specification to informal peer-group networks provides another interesting example of the potential strengths of the decentralized US standards system.[17] It reflects a fundamental distinction in standards development between system-level specification and component specification. While intense competition between leading global corporations dominates the process of system-level specification, component specification is outsourced to informal peer-group networks of engineers.

An example of system-level specification is the highly influential International Committee for Information Technology Standards (INCITS). INCITS is the primary US organization for creating and maintaining formal de jure standards in the field of information and communications technologies.

INCITS operates under rules, approved by ANSI, intended "to ensure that voluntary standards are developed by the consensus of directly and materially affected interests." Note, however, that INCITS is sponsored by the Information Technology Industry Council (ITIC), a trade association lobbying on behalf of "the world's leading innovation companies,"[18] most of them leading US providers of information technology products and services. It is informative to examine the INCITS executive board to identify who actually shapes strategic decisions. The INCITS executive board members encompass a "Who's Who" of US information technology companies, research labs and US government agencies (e.g., NIST and the Departments of Defense and Homeland Security).[19] According to interview sources, a small handful of companies, especially Intel and Microsoft, are the controlling members.

In its more than 50 technical committees, INCITS develops *system-level* specifications for the display, management, organization, processing, retrieval, storage, and transfer of information. These committees, however, only develop system-level specifications. INCITS does not attempt the tedious and time-consuming work requiring detailed feedback from customers to develop detailed component specifications. Developing detailed specifications requires extensive documentation—highly complex documents that are costly to generate and maintain.

Instead INCITS outsources detailed component specifications to specialized outside informal peer-group networks of engineers who work on these issues in member companies. One example is the Small Form Factor Special Interest Group (SFF-SIG), an independent non-profit industry group developing, promoting, and supporting detailed specifications for circuit boards and input/output and storage devices used in e-books, laptops, smart phones, and tablets.

SFF-SIG working groups are informal peer-group networks developed over time in this industry sector. Participants know each other and their individual interests, specializations, and strengths and generally complete their tasks without difficulties. According to one interview source, if problems arise, "you know exactly whom you need to talk to." Participant trust is critical in this tedious and challenging detailed specifications work. Equally important are the well-established relationships participants have with their customers and the continuing feedback the participants solicit from their customers.

Regular attendance at technical committee meetings is critical. Interview sources emphasize that, "If you join and you say you want to change something, you get absolutely no attention. You have to go to the meetings. If you only show up at critical meetings, no one takes you seriously…Informal networks can work out something fairly quickly. Most of the decisions are finalized between meetings or in the hall ways."

In short, the real strength of the US standards system are the multi-layered informal peer-group networks driving results in technical committees in organizations such as SFF-SIG.

What Are the Drawbacks of the American System?

This review of the US standards system has shown the extraordinary strengths of a market-led approach driven by the private sector. There are clearly valid rationales for China and other Asian countries to acknowledge and learn from the significant strengths of the deeply rooted US tradition of bottom-up, decentralized, informal, market-led approaches to standardization.

Every standards-development system however has strengths and weaknesses. Following are some of the weaknesses of the American system of voluntary standards.

Lack of Effective Coordination
The decentralized governance of the American standards system and its reliance on for-profit private firms comes with significant costs.

One such cost is the lack of effective coordination among the several hundred intensely competing private standards-development organizations that constitute the American standards system.

The American National Standards Institute (ANSI), a private-sector organization, was established in 1969 with the explicit mandate to "serve as a coordinator of the voluntary standardization aspect of... [the American standards]...system" (Hurwitz 2004).[20] But ANSI remains too weak. While formally the sole representative of US interests in international standards organizations, ANSI has been unable to reduce the intense rivalry among private standards organizations which continues to dominate the American standards system. What unites these private standard organizations is the "fear that a more centralized system would rob them of their revenues and eclipse their power and autonomy" (Mattli and Buethe 2003, 24).

ANSI's weakness is reflected by its limited involvement with congressional staff and US government agencies. ANSI has failed to attract the hundreds of consortia emerging in the information and communications technology (ICT) industry in part, according to Andrew Updegrove, "because of the reticence of these global organizations to appear more US centric than many of them are already perceived to be" (Updegrove 2008, 24).

The fragmentation of the US standards system is well documented. A study on "National Varieties of Standardization" finds that the US standards system is "by far the most institutionally heterogeneous and fragmented of all advanced industrialized countries" (Tate 2001). The lack of effective coordination by non-profit public actors may well produce negative results. The current US standards system "depends on consensus, negotiated among competing interests...[and] may lock in inferior technologies....Without public-interest representation,...special interests have powerful incentives to seek control of the process" (Alic 2009, 7–8).

> *The US standards system is 'by far the most institutionally heterogeneous and fragmented of all advanced industrialized countries'*

Intense Conflicts: The Battle Over Open Document Standards

The lack of effective coordination implies that the US standards system is prone to controversy and conflict. This does, in fact, describe the fundamental dilemma of the US system: its very strength—the diversity of stakeholders involved in standardization—also leads to intense competition and conflict among standards-development organizations and standards consortia, eroding the system's effectiveness and fairness.[21]

These concerns were substantiated in a now-classic study of the American standards system prepared by the Office of Technology Assessment (OTA) for the House Committee on Science, Space, and Technology of the US Congress. The study states that "concerns about the US standards setting process and recommendations for greater government involvement are based on the notion that the US approach no longer works as well as it should" (Office of Technology Assessment 1992, 7).

The OTA study emphasizes that the initial strengths of the American standards system have been a pragmatic bottom-up, decentralized, informal, market-led, standardization approach and the capacity to react swiftly to specific industry needs. Over time, however, the limitations of that system now outweigh its initial advantages. The US standards community is characterized by intense economic competition and personality conflicts. As a result, "internecine warfare in the standards community…raises questions about the ability of the voluntary standards organizations to carry out the public trust delegated to them" (Office of Technology Assessment 1992, 13).

A well-documented case of such "internecine warfare" resulting from the US-style governance structure was the battle to establish an international open document standard. This conflict pitted two competing file-format standards against each other—Microsoft's Office Open Extensible Markup Language (OOXML) versus the Open Document Format for Office Applications (ODF) developed by the Organization for the Advancement of Structured Information Standards (OASIS; of which International Business Machines [IBM], Oracle, and Sun Microsystems are strong supporters) Consortium. The selection on April 2, 2008, of OOXML as an ISO/IEC standard (ISO/IEC 29500) gave rise to an intense controversy. According to an editorial in the *Financial Times*, "Allegations of committee-stuffing, the outcome of votes overridden by political appointees, a final decision that many involved consider tainted: this may sound like a discredited election in some third world country. But it is actually a description of an ugly fight over international technical standards (i.e., the certification of Microsoft's OOXML standard by the International Organization for Standardization)."[22]

In an open letter, Nicos Tsilas, senior director of interoperability and intellectual property policy at Microsoft, attacked IBM's opposition to OOXML, arguing that IBM has led a global campaign urging national bodies to demand that ISO/IEC Joint Technical Committee 1 (JTC1) not even consider OOXML, because ODF had made it through ISO/IEC JTC1 first. According to Microsoft's Tsilas, IBM is "doing this because it is advancing their business model. Over 50 percent of IBM's revenues come from consulting services....[IBM is] using government intervention as a way to compete" as it "couldn't compete technically."[23]

In turn, Bob Sutor, vice president of standards and open source for IBM, criticized Microsoft's OOXML as "technically inferior...IBM believes that there is a revolution occurring in the IT industry, and that smart people around the world are demanding truly open standards developed in a collaborative, democratic way for the betterment of all,...If 'business as usual' means trying to foist a rushed, technically inferior and product-specific piece of work like OOXML on the IT industry, we're proud to stand with the tens of countries and thousands of individuals who are willing to fight against such bad behavior."[24]

The outcome of this fight was messy. According to the aforementioned *Financial Times* editorial, "Microsoft came out on top, but at the cost of tarnishing its reputation and the credibility of an important back-room process that oils the wheels of many global industries."[25]

The irony is that, after all the public conflict, Microsoft probably won a battle but cannot be sure to have won the war. Alex Brown, who had been the convener of the February 2008 JTC1 ballot-resolution meeting, in 2010 posted an entry on his personal blog in which he complained of Microsoft's lack of progress in adapting current and future versions of Microsoft Office to produce files in the "strict" (as opposed to the "transitional") ISO 29500 format: "On this count Microsoft seems set for failure. In its pre-release form, Office™ 2010 supports not the approved Strict variant of OOXML, but the very format the global community rejected in September 2007, and subsequently marked as not for use in new documents—the Transitional variant. Microsoft is behaving as if the JTC1 standardization process never happened" (Brown 2010).

Conflict Reduction: Suggestions of the 1992 OTA Study

To address the conflicts resulting from the decentralized governance of US standardization, the aforementioned 1992 OTA study suggested three strategies to reform and to upgrade the US standards system. OTA first suggested, based upon the "public-good" aspect of standards development, providing more government support for standards-development processes to address the failures of the market-driven system. OTA highlighted the coordinating role played by the national standards bodies of the UK (The British Standards Institution [BSI]) and Germany (Deutsches Institut für Normung e.V. [DIN], the "German Institute for Standardization") and deplored the current absence of a similar US organization. OTA also emphasized that US government agencies such as the Department of Commerce and the United States Trade Representative (USTR) respond to business queries and concerns ad hoc but that no agency has a mandate to develop a national standardization strategy. OTA argued that such a central coordinating agency is necessary to reap the potentially significant advantages of the voluntary standards system.

As a second strategy OTA suggested the development of a government "information infrastructure for accessing and distributing standards" and government participation in the standardization processes. OTA deplored America's standards infrastructure as a patchwork of mostly unconnected databases—most controlled by a handful of global industry leaders. According to the study, attempts to coordinate and extend the existing standards-information infrastructures were constrained by institutional inertia, resistance to changing the status quo, and a lack of the financial resources needed to make such changes.

The third strategy proposed by OTA was to improve the process of standardization through organizational restructuring. Such a strategy would have had to overcome deeply entrenched barriers: "Organizational arrangements are not neutral; they define power relationships determining who shall control what and for what ends" (Office of Technology Assessment 1992, 31).

Few, if any, of the OTA policy suggestions were implemented, despite many of the 1992 OTA findings still being relevant. Rather than gaining increased public attention, standardization concerns appear to have faded further into the background. With but a few exceptions it is difficult to find substantial discussion about standardization in public media.

> *Rather than gaining increased public attention, standardization concerns appear to have faded further into the background*

Shortfalls in the Provision of Strategic Standards

Another critique of the decentralized market-driven US system highlighted the potential tension between the dominant role currently played by for-profit firms in the governance of standardization processes and the role that standards are expected to perform in serving public-policy objectives. The expressed concern was that the dominance of private firms "may lead to consensus without…[adequate]…public-interest representation" (Alic 2009, 8). This may lead to market imperfections, such as the failure to adequately address public-policy objectives.

In the late 1980s a study by the MIT Commission on Industrial Productivity argued, in a chapter entitled "Failures of Cooperation," that a fundamental weakness of the US system was "the under-provision of such collective goods as joint research and development, standardization, education[,] and training, which…[are] instrumental in promoting technological innovation and productivity growth" (Dertouzos et al. 1989, 105).

Theoretical research on standardization may clarify why the US standards system might produce such failures. A fundamental insight from this research is that standards constitute a critical part of a country's economic infrastructure. Standards "help to determine the efficiency and the effectiveness of the economy; the cost, quality, and availability of products and services; and the state of the nation's health, safety[,] and quality of life" (Garcia 1993, 2). Standards reduce the costs and risks of market transactions and are necessary to reduce the costs of doing business.

Standards also provide the enabling infrastructure needed to enhance a country's innovation capacity. In a widely quoted book on the US innovation system, Greg Tassey (a senior NIST economist) argues that innovation requires a diversified and pervasive set of strategic standards—such as interoperability standards, security protocols, product specifications, and the formats and protocols that govern data transfer and interpretation (Tassey 2007). Tassey emphasizes that standards are a critical technical infrastructure that define the efficiency and effectiveness of a national innovation system. A broad portfolio of strategic standards is necessary to drive major innovations such as the "Smart Grid" project or the development of new alternative energy technologies.[26] Strategic standards are as important for a country's innovative capacity as are human capital, intellectual property rights, IT infrastructure, R&D investment, and venture capital. Underinvestment in strategic standards is as negative for growth as is underinvestment in education or in IT infrastructure. Innovation policy must therefore include the development of strategic standards as a key policy variable.

A broad portfolio of strategic standards is necessary to drive major innovations

Creating and maintaining these strategic standards requires extensive financial and human resources. When private interests dominate standardization, the "public good" nature of these strategic standards may well lead to a market failure, i.e., an under-provision of the necessary investments. Tassey deplores the US lack of a strategic vision integrating standards and innovation policy. He argues that, in the United States, the development of a technical infrastructure supporting innovation, and especially standards development, is "not receiving adequate levels of resources due to a poor understanding of such infrastructures' roles in long-term economic growth" (Tassey 2007, 240). Tassey concludes that the failure of policymakers to understand the complementary relationship between technology development and the development of supporting standards infrastructures is likely to erode US competitiveness.

To understand the negative impact of underinvestment in strategic standards, it is necessary to examine two fundamental changes in the nature of standards development:

- the growing importance of intellectual property management in standard-setting processes leading to sophisticated corporate strategies regarding "strategic patenting;" and
- the struggle to develop standard-setting processes that are fair, reasonable, and non-discriminatory and open to all stakeholders who are directly and materially affected by the standards—as manifested in the elusive concept of "open standards."

Strategic Patenting

As technology-based competition increases, a major key to competitive success is a broad portfolio of "essential patents" needed to produce any product that meets the specifications defined by the standard.[27] According to Ray Alderman, the chief executive officer of the VITA's (VITA, formerly the VMEbus International Trade Association's) Standards Organization (VSO), essential patents "are asserted against makers of products that are compliant to the standard.... The money is made through licenses and royalties, by asserting those claims against companies who, by implementing the requirements of the essential elements in the standard, must infringe the patent in order to make their products compliant to that standard" (Alderman 2009, 2–3).[28]

The use of "essential patents" as a strategic weapon to delay, obstruct, or prohibit standardization processes is well documented.[29] An example is when incumbent market leaders pursue "platform leadership" strategies through allegedly open but de facto proprietary standards.[30] While nominally "open," these standards are designed to block competitors and to deter new entrants to the market. Two highly influential studies by M.A. Lemley on the licensing and disclosure of private standards-setting organizations document the difficulties of finding fair, reasonable, and non-discriminatory patent-licensing-condition compromises in private standards-setting organizations (Lemley 2002; Lemley 2007).[31]

This is especially difficult for industries such as the information and communications technology sector where interoperability standards are required to make products or services compatible to maximize the benefits of network externalities. The Federal Reserve Bank of Philadelphia found this is made even more difficult by "the potential for opportunistic behavior by participants who own patents on a technology essential to the standard. There is a risk that without sufficient transparency and sufficiently strong mutual interests, network participants could make large investments to implement a standard only to be held up by a firm threatening to withhold a key piece of technology" (Hunt et al. 2007). The study argues that "in all likelihood some kind of agreement would be reached, but on terms substantially worse than the participants initially expected. Indeed, the risk of such an outcome may discourage firms from adopting a standard or even participating in the standard-setting process. In other instances, awareness of a key blocking patent might lead to the adoption of a standard that poses less risk to participants but which is also technologically inferior" (Hunt et al. 2007, 3).

The root of these negative outcomes may be found in market imperfections typical with contemporary high-tech industries. The emergence of a "winner-takes-all" competition model, described by Intel's Andy Grove, implies that companies need to combine economies of scale and scope with flexibility and speed-to-market (Grove 1996). Only those companies that succeed in bringing new products to relevant markets ahead of their competitors will thrive. Of critical importance is a firm's ability to build specialized capabilities more quickly and less expensively than its competitors (Kogut and Zander 1993). Competitive success critically depends on "a capacity to control open-but-owned architectural and interface standards" (Ernst 2002, 330). It is hardly surprising that, as John Alic puts it, under such conditions "firms may be tempted to seek profits through collusion rather than technological innovation. And when innovations do result, the costs may be high" (Alic 2009, 3).

As technology-based competition increases, a key to success is a broad portfolio of 'essential patents'

This has important policy implications. Lemley argues that the law must accommodate the way private standards-setting organizations (SSOs) deal with intellectual property. He argues that "antitrust rules may unduly restrict SSOs even when those organizations are serving pro-competitive ends. And enforcement of SSO intellectual property rules presents a number of important but unresolved problems of contract and intellectual property law, issues that will be needed to be resolved if SSO intellectual property rules are to fulfill their promise of solving patent holdup problems" (Lemley 2002, 1891).[32] However Lemley also warns against the danger of excessive regulation.

The following quotes from Lemley (2002, 1891–92), capture nicely the fundamental idea that underlies, at least in principle, the US standards system: "SSOs are a species of private ordering that may help solve one of the fundamental dilemmas of intellectual property law: the fact that intellectual property rights seem to promote innovation in some industries but harm innovation in others." Lemley is optimistic that SSOs will find ways to "ameliorate the problems of overlapping intellectual property rights in those industries in which intellectual property is most problematic for innovation, particularly in the semiconductor, software, and telecommunications fields."

For Lemley this implies that "*the best thing the government can do is to enforce these private ordering agreements and avoid unduly restricting SSOs by overzealous antitrust scrutiny*" [emphasis added].

In short, the use of "strategic patenting" to generate rents from de facto industry standards has transformed the dynamics of the US standards system. It has certainly made it more difficult to retain "open access" as a fundamental principle of the US standards system. Interviews with US standards engineers show their genuine commitment to a bottom-up, decentralized, informal, market-led, standardization approach and "open access." Unfortunately, however, the reality of standards consortia today is shaped by the race to squeeze profits out of the control of standards development.

This fundamental tension within the US standards system is well documented. For Branscomb and Kahin (1995), the main drivers of standards consortia are companies with large portfolios of essential patents. A fundamental weakness of the existing US standards system is that users (implementers as well as, especially, final users) continue to lack voice. This implies that the role of the government should not be restricted to that of being a user of standards. Equally important— yet clearly missing—is a sufficiently strong capacity for the US government to play the role of enabler and coordinator of standardization.

The Elusive Concept of "Open Standards"

Another critical weakness of the American voluntary standards system is the elusiveness of the concept of "open standards." Open standards have become almost an article of faith in the American standards system. Yet, according to the RAND Corporation's Martin Libicki, "all vendors pay lip service to open systems, but agreement ends here. The computer industry needs as many words for 'open' as Eskimos need for snow" (Libicki 1995, 43–44).

An in-depth RAND Corporation study on "Standards and Standards Policy for the Digital Economy" finds that "market leaders are rarely friendly to open standards when they dominate and eager to see them when they do not.... Market leaders are also friendly to standards in layers above and below them so as to use the competition among others to increase choices, lower costs, and broaden the market" (Libicki et al. 2000, 111).

For Libicki (1995, 42), the elusiveness of the concept of "open standards" implies that a neutral form of public governance is needed "to avoid the Scylla of chaos and the Charybdis of monopoly." Market-led standardization needs to be complemented by the US government to channel "the struggles of competing vendors and their technologies and the power of vendor versus user."

In principle this public governance role could be played by ANSI. According to the *ANSI Essential Requirements: Due Process Requirements for American National Standards,* standards developers accredited by ANSI must meet the institute's requirements for openness, balance, consensus, and other due process safeguards (ANSI 2010a).

Note, however, that ANSI narrowly defines "openness" as "a collaborative, balanced, and consensus-based approval process." According to ANSI's *Essential Requirements* document, "openness" means that "participation shall be open to all persons who are directly and materially affected by the activity in question. There shall be no undue financial barriers to participation. Voting membership on the consensus body shall not be conditional upon membership in any organization, nor unreasonably restricted on the basis of technical qualifications or other such requirements" (ANSI 2010a, 4). While this sounds good, the criteria used for measuring "openness" are much too abstract to work in the rough and messy world of intensive technology-based competition.

ANSI's narrow definition of "open standards" contrasts with the new benchmark for global open standards that five leading international-standards-development organizations—the IEEE, Internet Architecture Board (IAB), IETF, Internet Society, and World Wide

Web Consortium (W3C)—announced in August 2012.[33] The "shared open standard principles" draw on the effective and efficient standardization processes that have made the Internet and web the premiere platforms for innovation and borderless commerce and that have fostered competition and co-operation and supported innovation and interoperability across different layers of complex technology systems. This new approach to open standards

> *ANSI's narrow definition of 'open standards' contrasts with the new benchmarks for global open standards*

in based on five principles: a) cooperation among standards organizations; b) adherence to due process, broad consensus, transparency, balance, and openness in standards development; c) a commitment to technical merit, interoperability, competition, innovation, and benefit to humanity; d) availability of standards to all; and e) voluntary adoption (Mills 2012).[34]

For the United States to adjust to this new international norm of "open standards" it would be necessary to develop and strengthen cooperation between public and private actors in standardization. On the positive side, economic historians have shown that the United States has a long tradition of public-private partnership. According to David M. Hart, the concept of the "associative state" describes a fundamental characteristic of the US innovation system—the role of the state is to remedy "the informational failures of capitalism through cooperative inter-firm and business-government interaction" (Hart 1998, 420). Hart argues that, despite the twists and turns of antitrust policy and the rise and fall of the welfare state and the warfare state, a basic commitment to an "associative vision" of business-government relations has endured.

Unfortunately, however, the "deregulation wave" since the late 1970s has eroded the foundations of public-private partnerships. "Deregulation" is defined as the removal or simplification of government rules and regulations that constrain the operation of market forces (Derthick and Quirk 1985). In the United States, deregulation gained momentum on the back of theories from economists such as Friedrich von Hayek, Milton Friedman, and Ludwig von Mises, who argued that the economy was overregulated and that this imposed unnecessary costs on consumers (Baumol and Blinder 1991, 656).

As deregulation worked its way through the US economy it created a broad consensus among the Washington policymaking elite that actors in the private sector should be left free "to devise their own solutions to economic stagnation and international competition" (Russell 2006, 77). For standardization this implied that the role of the government should remain subdued and constrained to its function as a user of standards. There was limited opportunity for strengthening public-private interaction in the US standards system.

A particularly controversial issue is the implementation of the "voluntary consensus standards" central to the US standards system (Garcia et al. 2005). The National Technology Transfer and Advancement Act (NTTAA) of 1995 defines a "voluntary consensus standard" as a standard "that is developed through a process that entails 1) openness, 2) balance of interest, 3) due process, 4) an appeals process, and 5) consensus, defined as general agreement but not necessarily unanimity" (Office of Management and Budget 1998).[35] While this concept may look very attractive on paper, its implementation faces serious difficulties. It is thus problematic to use this concept as a "best practice" model for reforming the international standards system.

The pertinent section, 4(B), of the 1998 Office of Management and Budget (OMB) *Circular A-119* fails to establish a preference for "voluntary consensus standards." It explicitly allows for other private-sector standards to include "non-consensus standards," "industry standards," "company standards," or "de facto standards" that do not meet OMB's defining characteristics of "openness." Nor has the role and nature of consortia been addressed. These omissions have led to inconsistencies in the act's implementation. For example, while government agencies must report to NIST their progress in adopting voluntary consensus standards, consortia are not required to do so.

As policymakers have avoided these important issues, rivalries among standards-setting organizations have reduced many of the public benefits associated with voluntary standards.

The Role of the US Government

The usually unexamined "missing link" of the American standards system is the important role that the US government has played in shaping the evolution and the defining characteristics of that system.

The earlier-cited study by Libicki finds that "Protests to the contrary, the US government is a major, indeed increasingly involved, player in virtually every major standards controversy" (Libicki 1995, 35).

What role precisely has the US government played in fostering and shaping the US standards system? How has this governmental role evolved over time? Has the government's role in standardization helped to coordinate and channel the tremendous entrepreneurial and innovative energies that are set free in the pluralistic creation of voluntary consensus standards? What can be learned about the government's contribution to accomplishing the objectives of an innovation policy seeking to strengthen US innovation capabilities?

US Government's Direct Role: Standard-Setting Labs and de jure Standards

Direct government action involves the development of standards in government labs and the codification of mandatory standards requiring the use of specific standards through the force of laws or regulations.

In response to the establishment of national standard-setting laboratories in Britain and Germany, in 1901 the US Congress created the US Bureau of Standards. The initial mandate was to coordinate the rapid proliferation of scientific standards as well as to carry out scientific research in its own laboratories (Cochrane 1966). Initially the Bureau of Standards focused its efforts narrowly on standards for heat, optics, measures, and weights. Over time the Bureau of Standards (which changed its name to the National Institute for Standards and Technology [NIST] in 1988) expanded its mission to include electricity research as well as testing of materials quality and also provided technical assistance and product evaluations for regulatory bodies. But never has it played a role at all comparable to the German Institute for Standardization (DIN) in coordinating, implementing, and shaping the nation's standards strategy and policies.[36]

As shown earlier, ANSI has a much more limited mandate. Its primary objective is to represent the interests of its nearly one thousand members, most of them private companies. ANSI's role is restricted to "promoting and facilitating voluntary consensus standards and conformity assessment systems and promoting their integrity…by accrediting the procedures of…[about 200 independent]…standards developing organizations…Accreditation by ANSI signifies that the procedures used by the standards body in connection with the development of American National Standards meet the Institute's essential requirements for openness, balance, consensus[,] and due process" (ANSI 2012).

Until World War II, the direct role of the US government has remained limited. De jure or regulatory standards were typically restricted to health and safety issues (e.g., the Pure Food and Drug Act of 1906 and the Meat Inspection Act of 1906) and the prevention of the abuse of market power (e.g., the Federal Commission Trade Act of 1914).

The government's direct role in the US innovation system however received a big push once the US entered the Second World War. The perceived threat from the Soviets during the Cold War added further momentum to a more activist role of the government. This gave rise to substantial investments by the Federal Government in the development of basic standards, such as the development of programming languages, measurement standards for optical fibers, computer-aided design technologies, and the basic Internet standard TCP/IP.

However, the deregulation wave that gathered momentum during the Reagan administration reversed this trend, substantially limiting the direct role of the government in standardization.

US Government's Indirect Role: The Impact of Antitrust Policy

An important finding of innovation research is that US antitrust policy has played an important role in the development and rapid diffusion of standards in US industry. Opinions, however, remain deeply divided on the pros and cons of aggressive versus passive antitrust policies.

A fundamental premise of the US standards system is that deregulation and the promotion of market competition are necessary to reduce "policy imperfections" generated by incompetent bureaucracies perceived to stifle innovation and productivity growth. Voluntary standards developed within informal consortia are believed best suited to solve "collective action" problems which prevent "rational, self-interested individuals…[from]…act[ing] to achieve their common or group interests" (Olson 1971, 2).

This dominant consensus is now under pressure as the global economic crisis beginning in 2007 again demonstrated the limits of deregulated markets. Today there is a greater willingness in Washington, DC, to revisit the merits of more activist anti-trust policies and regulations. Research by leading US innovation economists has demonstrated that, when handled appropriately, antitrust policy can be a powerful enabler of innovation and standardization.

The 1956 consent decree, resulting from antitrust pressures from the Department of Justice (DOJ), ordering the compulsory licensing of roughly 8,600 AT&T (formerly the American Telephone and Telegraph Corporation) patents and a nearly simultaneous decree affecting IBM patents both inspired intense public scrutiny.

Frederic M. Scherer (a leading innovation economist at Harvard University) finds that these decrees generated a "profoundly surprising" positive effect for "small new enterprises seeking a competitive foothold against well-entrenched rivals" (Scherer 1977; Scherer 2006, 5–6). By enabling small start-up companies to gain access to technological advances the consent degree provisions for compulsory licensing of AT&T and IBM patents arguably have been a powerful catalyst for the development of Silicon Valley start-up companies.

This finding is supported in the comprehensive and now-classic study *Sources of Industrial Leadership* by the University of California at Berkeley's David Mowery and Columbia University's Richard Nelson. Their study emphasized the important positive role of active US postwar antitrust policy:

> Although it rarely receives extensive attention in discussions of technology and competitiveness, the relatively stringent postwar competition policy of the United States aided the growth of new industries. US antitrust policy weakened the ability of incumbents in such industries as computers and semiconductors to control new technologies and markets.... [due to]…a relatively weak intellectual property rights environment for most of the first three decades of the US industry's development. (Mowery and Nelson 1999, 379–80)[37]

Testing the Limits: The US Department of Justice Supports VITA's ex ante Disclosure of Essential Patents

A more recent example of the potentially important role that US antitrust policy could play for the US standards and innovation system was the October 2006 decision by the DOJ to support a proposed patent policy by the VITA standards-development organization that requires ex ante disclosure of essential patents and their licensing terms.[38]

In an October 30, 2006, letter to the attorney representing VITA, the Assistant Attorney General states that the DOJ "has no present intention to take antitrust enforcement action against the conduct you have described." Specifically the letter states:

> Once a particular technology is chosen and the standard is developed, however, it can be extremely expensive or even impossible to substitute one technology for another. In most cases, the entire standard-setting process would have to be repeated to develop an alternative standard around a different technology. Thus, those seeking to implement a given standard may be willing to license a patented technology included in the standard on more onerous terms than they would have been prior to the standard's adoption in order to avoid the expense and delay of developing a new standard around a different technology.
>
> Requiring patent holders to disclose their most restrictive licensing terms in advance could help avoid this outcome by preserving the benefits of competition between alternative technologies that exist during the standard-setting process....
>
> The disclosure of each patent holder's most restrictive licensing terms would allow working group members to evaluate substitute technologies on both technical merit and licensing terms. Working group members are likely to use this information when deciding which technologies to include in the standard. This use likely will create incentives for each patent holder to compete by submitting declarations that will increase the chances that its patented technology will be selected....
>
> Adopting this policy is a sensible effort by VITA to address a problem that is created by the standard-setting process itself. Implementation of the proposed policy should preserve, not restrict, competition among patent holders. Any attempt by VITA or VSO members to use the declaration process as a cover for price-fixing of downstream goods or to rig bids among patent holders, however, would be summarily condemned. (Department of Justice 2006, 6–8)

The above decision from the DOJ constitutes an important change in the department's approach to the standard-setting processes. Until this decision the prevailing assumption was that collaborative standard-setting could result in exclusionary and collusive practices that would harm competition and violate US antitrust laws. That earlier understanding had led many SDOs to implement rules strictly forbidding any activities potentially resulting in antitrust liability, including restrictions on the discussions concerning terms and conditions of licenses to patents that are essential to a standard.

The DOJ opinion regarding VITA's proposed policy on ex ante disclosure helps to avoid any such unintended negative consequences and should motivate other SDOs to also gradually relax their similar restrictions. The following summarizes the DOJ's new position:

> Unless the standard-setting process is used as a sham to cloak naked price-fixing or bid rigging, the Department analyzes action during the standard-setting process under the rule of reason. The Department's analysis of VITA's proposed patent policy under the rule of reason examines both the policy's expected competitive benefits and its potential to restrain competition. (Department of Justice 2006, 6)

The DOJ's decision and VITA's implementation of this new patent policy has generated a continuing controversy. As summarized by the DOJ letter, the expected benefits of ex ante disclosure are substantial. Yet leading global IT companies have raised strong opposition.

Opponents argue that ex ante disclosure will have disruptive effects on the smooth functioning of the US standardization process and that it will stifle innovation. Opponents claim that the inherent uncertainty of technical change prevents correct and timely disclosure or would require extensive patent searches at very high cost. An additional critique is that important companies with large patent portfolios are unlikely to accept ex ante disclosure and hence would leave any SDO seeking to implement such a policy.

The expected benefits of ex ante disclosure are substantial. Yet leading global IT companies have raised strong opposition

Motorola, in fact, left VITA in protest against the new ex ante policy. More than 20 new companies (including Boeing, General Dynamics, General Electric [GE], Northrop Grumman, and Ratheon), however, have joined VITA since the new patent policy was established.

To express its fundamental opposition to the policy of ex ante disclosure Motorola also filed an appeal against the decision by ANSI's Executive Standards Council to reaccredit VITA. This appeal was dismissed by ANSI's Appeals Board Panel.[39]

In the academic literature the claims of "ex ante disclosure" opponents have largely met with skepticism. A recent PhD thesis by Claudia G. Tapia examines these and other related arguments against ex ante disclosure and concludes that "it is questionable whether the current skepticism towards ex ante disclosure is really justifiable." But she also adds that, "without further in-depth analysis, it remains unclear whether the mandatory process works and, if so, under what circumstances" (Tapia 2010, 198).

And a recent study prepared for NIST concluded that "the information elicited by the organization's ex ante policy was important and improved the overall openness and transparency of the standards-development process. Thus,…the process-based criticisms of ex ante policies and the predicted negative effects flowing from the adoption of such polices, are not supported by the evidence reviewed" (Contreras 2011, 1).

The main problem, in the critically important IT industry, seems to be that the strength of the opposition is such that, until now, no other US standards-development organization has decided to follow VITA's example. Despite the potentially substantial benefits of ex ante disclosure policies, opponents have succeeded in preventing the general acceptance of this policy.

VITA's main focus now is on the aerospace and defense industries. Adopting ex ante mandatory disclosure policies is possible in those industries as, given the strict procurement requirements of the US military system, these companies can easily afford to not pursue a "pure intellectual property" business model.

In contrast, the "pure intellectual property" business model continues to shape competition in large globalized industries (e.g., smart phones, integrated circuits, and telecommunications) that are scale-intensive, that depend on venture capital and private equity, and where speed-to-market is of the essence. In these industries competition is driven by "winner-takes-all" strategies and management must squeeze profits out of every stage of the value chain, including intellectual property rights and the standards process. Market inefficiencies in these industries are pervasive and systemic. These market inefficiencies constrain innovation and the supply of necessary innovation infrastructure (e.g., interoperability standards) and they obstruct the normal workings of the market.

> *Despite the potentially substantial benefits of ex ante disclosure policies, opponents have succeeded in preventing the general acceptance of this policy*

VITA's experience shows that standardization processes must be context-specific, i.e., they must take into account the structure and competitive dynamics of specific industries and market segments. There is no one best approach to establish a transparent intellectual property-rights policy and an open-standards system.

US Government Fails to be an Effective Coordinator

A defining characteristic of the US standards system is the limited ability of the US government to serve as a coordinator, enabler, or, if necessary, enforcer of rules to prevent possible abuse of market power by companies with extensive patent portfolios.

An influential 1995 study of the US standards system concluded that the development of standards for complex technology systems requires a division of labor between the government (in the role of coordinator and enabler) and the private sector (in the role of innovator and investor). If either of these two complementary elements is missing it will be difficult to generate the appropriate balance between public and private interests (Branscomb and Kahin, 1995)

The engagement of the private sector (as innovators and investors) in the American standards system represents a unique strength—standards are not imposed by the government, but "are expected to emerge from the experimentation, competition, and…the market response to the standards process and its expressions—reference models, architectures, draft specifications, or standards—and in the further response of the standards process to the market" (Branscomb and

Kahin 1995, 4). In an ideal world, where economic power is equally distributed among stakeholders in standardization, a private-sector-driven standards system might be effective in unlocking barriers to innovation.

In the real world, however, standardization is a contested and constantly changing field. Technology-based competition is intensifying with the result that standards are used everywhere to create and shape markets and to control competition. This requires a stronger government role as a coordinator, a provider of strategic vision, and a repository of knowledge. In the United States the federal government largely lacks the mandate and the resources to provide these three fundamental public services.

The deregulation wave since the late 1970s has reduced again the government to a largely passive role and allowed private industry to lead the way. Absent effective government coordination, for-profit corporations typically concentrate their efforts on advancing their proprietary technologies addressing specific problems—without regard as to whether or not these technologies provide systemic solutions. This leads to many standards often lacking in sufficient interoperability. This is a poor approach for the complex, large-scale technology systems typical of the information technology industry. IT involves multiple layers of standards needing to be identified, harmonized, and, importantly, broadly diffused to a diverse community of standards implementers and users.

The critical question is whether necessary adjustments in the balance of public-private interests can be implemented in time to allow the US standards system to cope with the global challenges of rapid innovation and globalization.

A New Approach to Public-Private Standardization Partnerships? The Smart Grid Interoperability Standards Project

The Smart Grid interoperability standards project,[40] coordinated by NIST, offers an important example of recent US attempts to move beyond the legacy of deregulation and search for new approaches to public-private standardization partnerships.

The Challenge of Rising Complexity

This project faces a daunting task. America's electricity grid is "aging, inefficient, and congested, and incapable of meeting the future energy needs of the Information Economy without operational changes and substantial capital investment over the next several decades" (Department of Energy 2003).

The Smart Grid "utilizes advanced information and communications technologies to enable a two-way flow of electricity and in-

'A 21st century clean energy economy demands a 21st century electric grid'

formation…to make the grid more efficient by reducing demand peaks and increasing capacity utilization and providing consumers with tools to reduce energy usage and potentially save money" (Arnold 2011).

The challenge is momentous, as described in a recent strategy document for the Smart Grid: "A 21st century clean energy economy demands a 21st century electric grid. Much of the traditional electricity infrastructure has changed little from the design and form of the electric grid as envisioned by Thomas Edison and George Westinghouse at the end of the 19th century" (NIST 2012).

The Energy Independence and Security Act (EISA) of 2007 made it official US policy to modernize the nation's electricity distribution and transmission system to create a smart electric grid (Energy Independence and Security Act [EISA] of 2007). A June 2011 report by the White House National Science and Technology Council laid out the strategy (National Science and Technology Council 2011).

George Arnold, the NIST coordinator for Smart Grid interoperability, believes nothing less than a complete transition "from today's electric grid, in which there has been a tradition of proprietary interfaces and product customization for individual utilities, to an interoperable grid based on open standards [is needed]…[This]…is a huge change for the industry" (quoted in Updegrove 2009; see also Arnold 2011). Essential to the success of this task will be the accommodation of traditional, centralized, generation and distribution resources while also facilitating the incorporation of new, innovative, Smart Grid technologies (such as distributed renewable energy resources and energy storage).

The *NIST Framework and Roadmap for Smart Grid Interoperability Standards* describes an unprecedented standardization challenge

(NIST 2012). Upgrading the existing patchwork of the North American power-system grid will require more than 75 existing major standards to be reviewed, adjusted, and approved so that they will work together. Hundreds of new requirements, specifications, and standards need to be created in such diverse fields as advanced-metering infrastructure, cyber security, distribution-grid management, electric transportation, energy efficiency, energy storage, and network communications to master the transition to the Smart Grid.

Increasing complexity in the Smart Grid project results from the inherent limitations of the existing disparate and uncoordinated networks. There are roughly 3,100 utilities in the United States involved in the power-system grid and more than 15 standards-development organizations. This is vastly different from the late 1970s ownership that automated the telecommunications network. At that time the entire US telephone network was owned by one company—AT&T. Planning and standards setting was far easier—it was all done by AT&T's Bell Labs.

The Smart Grid interoperability standards also need to respond to a complex regulatory environment. Beyond the federal government they must address an additional 51 jurisdictions (50 states plus the District of Columbia). Adding to the difficulties the project must quickly establish effective cooperation between two industries whose business models and strategies could hardly be more different.

The utility industry moves glacially, in part because of its complex regulatory environment. But an equal cause for the slow pace of change in this industry would arguably be its fragmented ownership structure.

Contrast this slow-paced rate of change with that of the providers of the information hardware and software to be used for integrating the new grid. While also numerous, these businesses are in the fast-moving IT industry where profits depend on speed as well as on strategic patenting.

Increasing system complexity greatly increases the difficulty in developing interoperability standards

Increasing system complexity greatly increases the difficulty in developing interoperability standards. Interoperability standards will be required for both interfaces among technology domains (e.g., between cyber security and distribution-grid management) and interfaces among different participants (primarily private firms from the multiple industries involved in the construction of the Smart Grid). This requires that the interoperability framework must be "flexible, uniform, and technology neutral" (NIST 2010, 7).

A Pragmatic Approach

The Smart Grid is conceived of as a "complex system of systems for which a common understanding of its major building blocks and how they interrelate must be broadly shared. NIST has developed a conceptual architectural reference model to facilitate this shared view. This model provides a means to analyze use cases, identify interfaces for which interoperability standards are needed, and to facilitate development of a cyber security strategy" (NIST 2010, 8).

NIST stipulates that

interoperability standards...[for the Smart Grid] should be open. This means that the standards should be developed and maintained through a collaborative, consensus-driven process that is open to participation by all relevant and materially affected parties and not dominated by, or under the control of, a single organization or group of organizations. As important, the standards resulting from this process should be readily and reasonably available to all for Smart Grid applications. In addition, Smart Grid interoperability standards should be developed and implemented internationally, whenever practical. (NIST 2010, 9)

NIST believes that the key to success is a pragmatic approach using whatever works best and discarding suggestions that do not quickly deliver technically sound open standards. As emphasized by George Arnold:

We are trying to do something with the grid that has not been done before. The interoperability in the telecommunications network is done almost entirely through voluntary standards, and it seems to work. However the electric grid is much more fragmented...and has more a tradition of using proprietary systems....[Hence]...*some combination of voluntary and mandatory standards will likely be needed* [emphasis added]." (quoted in Updegrove 2009, 6)

NIST argues that today this more flexible approach to standardization is made possible by the ubiquitous use of software already embedded in many important standards. New approaches to programmable system-on-chip devices make it possible to continuously update such equipment.[41]

Still the development of the Smart Grid faces tremendous time pressure. Attempting to achieve quick results within a context of high complexity increases the need for "open" interoperability standards.

Developing such open standards will only be possible if new forms of public-private standards-development partnerships are created. NIST expects the process of developing the Smart Grid will be a catalyst for developing "new collaborative methods and vehicles for developing and deploying standards in technology-based markets, especially during the early phases when standards—or the lack of standards—can strongly influence the course of further technological development and diffusion and the growth and competitiveness of industries" (NIST 2010, 11).

NIST considers the Smart Grid project an important experimentation opportunity to develop new governance mechanisms and methods for public-private standards-development partnerships.

Multiple Stakeholders with Conflicting Interests

Implementing this concept will not be easy. The Smart Grid project has attracted an extraordinary number of diverse organizations all seeking to shape and profit from Smart Grid interoperability standards.[42]

The critical role of government agencies. The Smart Grid interoperability standards project is currently distinguished by the prominent role being played by government agencies in shaping its agenda and in providing key resources and controlling project outcomes. Under EISA, the Department of Energy (DOE) has overall responsibility for the Smart Grid project while NIST is to coordinate the development of Smart Grid standards and is responsible for cyber-security. Both NIST and the DOE's Office of Electricity Delivery and Energy Reliability (OEDER) must report, on a regular basis, to congress regarding the status of Smart Grid deployments and any regulatory or governmental barriers to continued deployment (Congressional Research Service 2007). The Department of Homeland Security has been tasked with monitoring the Smart Grid for security against cyber attack.

Private-sector organizations. A wide range of private-sector SSOs are active in the creation of standards relevant to the Smart Grid. Some of the most prominent organizations developing key standards include the IEC, IEEE, IETF, National Electrical Manufacturers Association (NEMA), and the North American Electric Reliability Corporation (NERC). There are about 15 organizations in total including the consortium that develops ZigBee. ANSI does not play an overly prominent role. As George Arnold diplomatically phrased, "ANSI also [sic] has a key role in ensuring there is a good process for standards development and facilitating access to IEC and ISO" (quoted in Updegrove 2009, 4).

What is significant is that, at least nominally, all private-sector players (even the most powerful) have agreed to accept the coordinating function of NIST. For standards that effort is lead by George Arnold, the United States' first National Coordinator for Smart Grid Interoperability. Arnold, who was appointed to this position in April 2009, is well-respected within the standards community. He was formerly a vice president at Lucent Technologies' Bell Laboratories and was active in the development of international standards for intelligent networks and IP-based next-generation networks.

The appointment of George Arnold highlights an important strength of the American standards system. Not only do standards associations have a long history of independence, they can draw on a large pool of well-educated and -experienced standards experts who have developed their own peer-group networks. In the American system these individuals are often decisive as coordinators and gatekeepers in shaping decisions on standardization and in implementing the resulting standards.

At least initially, private-sector organizations will play a secondary role in the design and implementation of the Smart Grid project. Private-sector organizations are, presumably, accepting this subordinate role in expectation of reaping the benefits of the substantial externalities from tax-financed public investment in the required project infrastructure, support institutions, and R&D. An additional rationale for the accommodations of private-sector organizations may be found in the large project budget made available as part of the American Recovery and Reinvestment Act (ARRA) of 2009.[43]

By 2012, however, the 2009 stimulus-package funds had all been spent and venture-capital investment in the Smart Grid had tumbled.[44] It remains an open question whether this loss of Smart Grid funding and the resulting pressure on profits will create a more adversarial climate between private-sector organizations and government agencies.

Governance: The Smart Grid Interoperability Panel

To cope with these conflicting interests, NIST established the Smart Grid Interoperability Panel (SGIP) as the primary governing body for the development of Smart Grid interoperability standards. Made up of more than 450 standards organizations, utilities, vendors, and other related companies, SGIP has been tasked with performing interoperability tests on the 25 approved standards as well as attempting to resolve any conflicts or problems in the remaining 50 standards not yet been approved. The main tasks of SGIP are:

- to provide a more permanent process with stakeholder representation in order to support the ongoing evolution of the Smart Grid Interoperability framework;
- to identify and address additional gaps [and] reflect changes in technology and requirements in the standards;
- and to provide ongoing coordination of SSO efforts to support timely availability of new or revised Smart Grid standards (NIST 2010, 116).

As specified in EISA, the SGIP governing board is "an open, transparent public-private partnership to support NIST in its primary responsibility to coordinate development of a framework that includes protocols and model standards for information management to achieve interoperability of Smart Grid devices and systems." To maintain a broad perspective on the NIST interoperability framework, the SGIP governing board is responsible for approving and prioritizing the work of the SGIP and coordinating necessary resources to effectively implement action plans.

To help ensure that all stakeholder categories are fairly represented on the SGIP governing board, members must have extensive experience in one or more stakeholder categories and the ability to support overall SGIP and NIST goals. Current SGIP governing board members have been selected from over 23 stakeholder categories including utilities, renewable power producers, standards development organizations, professional societies, manufacturers and vendors, consumers, and IT and system developers.

SGIP does not directly develop or write standards; rather its stakeholders participate in the ongoing acceleration, coordination, and harmonization of standards development. Key corporate members include ABB, Ford Motor Company, GE Energy, Google, Honeywell, and Lockheed Martin and industry associations such as NEMA.

An Unresolved Issue: How to Manage Essential Patents for Smart Grid Standards

A critical challenge for the successful implementation of the Smart Grid standardization project is the management of the intellectual property (especially patents) essential for the required standards. Given the rapid growth and innovation in information technology, many components of the Smart Grid are or will be covered by patents.

An analysis of Smart Grid patent data found 2,469 company-owned patents spread among 697 companies. However 39 companies, each owning at least 10 Smart Grid patents, clearly dominated the listings (Fisher and Nirenberg 2010). Vendors of equipment and software such as ABB, GE, Hitachi, Samsung, and Siemens are the dominant owners of Smart Grid patents. "These same vendors will also use that intellectual property to compete for leadership in sales to the utilities, who mainly want the benefits from Smart Grid developments without the cost of associated research" (Fisher and Nirenberg 2010, 7).

A critical challenge for the successful implementation of Smart Grid standardization is the management of intellectual property—especially patents

The unequal distribution of Smart Grid patents could reduce the benefits of Smart Grid standardization. According to a recent in-depth study of patent management in the Smart Grid project,

If patents that cover standardized Smart Grid elements are not revealed until technology is broadly distributed through-out the network ("locked-in"), significant disruption could oc-cur if patent holders sought to collect unanticipated rents...As a result, costs to consumers could increase, competitors could be shut out from the market, and the standardization process itself could be subverted....[For Smart Grid standardization, this risk is even higher than for information and communica-tions technology]...as Smart Grid standards are mandated by law and have the potential to be adopted into both federal and state regulation, making lock-in nearly impossible to avoid and providing even greater leverage to opportunistic patent holders. (Contreras 2012b, 642–43)

This returns attention to a fundamental weakness of the American standards system, the reluctance to implement effective mechanisms to avoid misuse of patent ownership for standards development. Nei-ther NIST nor SGIP have thus far established mechanisms to reduce opportunistic enforcement of patents covering key Smart Grid stan-dards. Contreras argues that the government must correct this impor-tant failure: "Thus, in order to ensure the rapid deployment and unin-terrupted operation of the national Smart Grid, it is incumbent upon NIST, [Federal Energy Regulatory Commission] FERC, and Congress to implement rules that will maximize transparency of the standards-development process and prevent disruption of this critical national resource" (Contreras 2012b, 675).

Prospects

It is too early to tell whether the model of joint public-private stan-dardization partnership provides a robust framework for solving the daunting tasks of Smart Grid interoperability. Speed and efficiency it may well improve—but it remains an open question whether this model will provide a fair distribution of the costs required and the rents to be reaped from Smart Grid standardization.

While ample evidence points to the need for the stronger role of smart regulations on and a greater scrutiny of essential patents, the decentralized, market-led approach to standardization appears, for now, to retain the upper hand. Thorough empirical follow-up research is needed that examines the effects of this important experiment in large-scale public-private standardization partnership.

Conclusion

The American approach to standardization reflects the unique characteristics of its history and economic institutions. Thus, even if the strengths outweigh the weaknesses of the US standards system—which this study argues is the case—it would still be difficult to fully replicate the US system in other countries. This is especially true for Asian countries with their very different economic institutions.

The flexibility and bottom-up character of the US market-led system of voluntary standards has been an important source of America's extraordinary record in generating commercially successful innovations. Grounded in a tradition of decentralized local self-government, the American standards system has given voice to diverse innovation stakeholders and has avoided the pitfalls of top-down government-centered standards systems. Such government-centered systems are hard to start, harder to adjust or steer, and still harder to stop once they get going.

Examples of the strengths of the American system discussed here are the IETF model of system-level standardization for the Internet and, especially in the IT industry, arrangements for outsourcing of component-specification tasks.

For Asian countries seeking to upgrade their standards systems it makes sense to study the strengths of the American system. However this study documents, as well, significant limitations and weaknesses of the US market-led system of voluntary standards. Especially in the management of essential patents and in the timely provision of interoperability standards, intense conflicts and a lack of effective coordination of multiple-stakeholder strategies have created serious constraints to effective and open standardization processes.

> *The American standards system has given voice to diverse innovation stakeholders*

As standardization is a contested and constantly changing field of economic activity with far-reaching implications for competition and market structure, the government must play an important role as a coordinator, an enabler, and, when necessary, an enforcer of the rules of the game in order to prevent excessive abuse of market power by companies with large patent portfolios. Globalization and the increasing complexity of advanced technology imply that public policy must balance and complement the strengths of the market-led approach to standardization. This study shows this to be true for the United States, where the strengths of a decentralized, voluntary standards system need to be balanced and complemented by strong coordination mechanisms and smart regulations, especially in such contested areas as standards-essential patents.

Globalization and the increasing complexity of advanced technology imply that public policy must balance and complement the strengths of the market-led approach to standardization

To cope with the new challenges of the global knowledge economy, America's standards system needs to be reshaped and upgraded as part of an integrated innovation policy. The US government and the private sector need to join forces and develop a national innovation strategy seeking to combine productivity improvements with the creation of quality jobs while minimizing energy usage, materials waste, and other environmental impacts (Ernst 2012).

A closer look at the efforts of the US government shows that, while working largely away from public attention, it has frequently been an active participant in standards development. Even though direct government action through standard-setting labs and de jure standards remains limited, primarily through antitrust policies, the government has played an important role. This study argues that, after an extended period of passive antitrust policies, the pendulum now appears to be swinging back to a more activist approach. This is illustrated by the DOJ's support of ex ante disclosure of essential patents.

Of great interest for observers from Asian countries, and especially from China, are the current attempts to establish and strengthen robust public-private standards-development partnerships. To examine this important development, this study has analyzed in detail the US Smart Grid interoperability standards project coordinated by NIST. This project is currently distinguished by the prominent role being played by government agencies in shaping its agenda and in providing key resources and controlling project outcomes. This offers an interesting comparison with China's approach to Smart Grid standardization (Ziegler 2011; see also State Grid Corporation of China 2011).

In conclusion this study identifies significant considerations for standards and innovation policies in Asia. Attempts to copy and replicate the US standards system will face clear limitations. While standards systems everywhere are confronted with similar tasks, there are significant differences in the organization and governance of standardization processes. These differences reflect the unique characteristics of each country's differing economic institutions, their levels of development, their economic-growth models, and their cultures and history.

Endnotes

1. This study is a draft of a book chapter for Ernst, D., forthcoming, *Innovation Policy in the Global Knowledge Economy: Comparing the US, China, Taiwan, Korea and Europe*. An earlier version of this study has been published as an East-West Center Economics Working Paper. This study is the counterpart to an earlier book on China's standards system (Ernst, D. 2011. *Indigenous Innovation and Globalization: The Challenge for China's Standardization Strategy*. La Jolla, CA: UC Institute on Global Conflict and Cooperation and Honolulu: East-West Center. http://www.EastWestCenter.org/pubs/3904 [published in Chinese by the University of International Business and Economics Press, Beijing])

2. These interviews are part of the East-West Center's research on innovation policy in the global knowledge economy.

3. For recent research on the barriers to open access to SDOs, see Contreras, 2012a, and Updegrove, 2012. See also the pioneering research reported in Simcoe, 2012.

4. *The Economist*, February 27, 1993, "Do It My Way," 11–12.

5. This assessment that China gave away too much in joining the World Trade Organization (WTO) is supported by Alan Wm. Wolff who argues that "In 2001 China joined the WTO. In doing so, China made a much more extensive number of specific market liberalizing commitments than had ever been made by any acceding member, and began to transform its economy" (Wolff 2010, 3).

6. There are, of course, many specialized data bases for engineers that compare technical standards for particular technologies. But very little research exists that compares institutional arrangements and strategies that shape different national standards systems.

7. Toru Yamauchi is the former Director of Industrial Standards Research, Japan International Cooperation Agency.

8. An example of this outdated view of the global map of national standards systems can be found in Mattli and Buethe (2003). See, however, a new project by the National Academy of Sciences seeking to compare different national systems of

managing intellectual property in standards-development organizations. http://sites.nationalacademies.org/PGA/step/IPManagement/index.htm.

9. See the analysis of "first principles" of successful standardization in Contreras 2012a.

10. Fortunately the recent surge in academic interest in standardization is no longer limited to research on business, economics, and engineering. There is now a substantial body of research on the historical development of standardization. See, for instance, Russell 2005.

11. According to a contemporary New York journalist, the successful standardization of the nation's railways demonstrated beyond doubt that "the laws of trade and the instinct of self preservation effect reforms and improvements that all legislative bodies combined could not achieve" (quoted in Kirkland 1961, 50).

12. http://www.whitehouse.gov/sites/default/files/omb/memoranda/2012/m-12-08.pdf.

13. The author is a former standards manager at Oracle. His interview sample included key players of the Standard Policy Committee of the influential Information Technology Industry Council.

14. In computer-network engineering, an Internet standard is a normative specification of a technology or methodology applicable to the Internet.

15. The World Wide Web Consortium (W3C) is the main international standards organization for the World Wide Web. The International Organization for Standardization (ISO) is an international-standard-setting body, based in Geneva, composed of representatives from various national standards organizations. The International Electrotechnical Commission (IEC) is a non-profit, non-governmental, international standards organization that prepares and publishes international standards for electrical, electronic, and related technologies.

16. The Internet protocol suite (commonly identified as TCP/IP) is a set of communications protocols used for the Internet and other similar networks. It includes two of the most important protocols: the transmission control protocol (TCP) and the Internet protocol (IP), the first two networking protocols defined by these standards.

17. This is based on interviews conducted September 8 and 23, 2009, with a number of US standards experts who requested anonymity.

18. http://www.itic.org/about/

19. Key corporate members of the INCITS executive board include Adobe, Apple, Google, Hewlett-Packard, IBM, Intel, Lexmark, Microsoft, and Oracle; see http://www.incits.org/ebmem.htm.

20. Mr. Hurwitz was then president and chief executive officer of the American National Standards Institute.

21. I am grateful for information provided by participants in the development of the international open document standards who have requested to remain anonymous.

22. *Financial Times*, April 3, 2008, Editorial.

23. Quoted in ZDNET, http://www.zdnet.com/microsoft-ibm-masterminded-ooxml-failure-3039292492/

24. Quoted in Ars Technica, http://arstechnica.com/microsoft/news/2008/02/ibm-responds-to-microsoft-ooxml-is-technically-inferior.ars, accessed May 6, 2010.

25. *Financial Times*, April 3, 2008, Editorial.

26. Tassey uses the term "infratechnology" standards. I prefer to call them "strategic standards," highlighting their critical importance for upgrading a national innovation system.

27. Patents are "essential" to a standard when it is not possible to comply with the standard without infringing on intellectual property rights.

28. For a description of VITA's patent policy see "Testing the Limits: The US Department of Justice Supports VITA's ex ante Disclosure of Essential Patents" in this study.

29. See the seminal article by Lemley and Shapiro (2007). For an analysis of implications for standards-development organizations and policymakers see Weiss and Spring (2000).

30. The overriding purpose of "platform leadership" strategies is to leverage the existing market power of industry leaders into the control of "systemic architectural innovations" (see Gawer and Cusumano 2002, 39). For example, Intel has attempted to extend its control over microprocessors by creating widely accepted architectural designs that increase the processing requirements of electronic systems and, hence, the market for Intel's microprocessors (Gawer and Henderson, 2007).

31. See also the recent systematic study by Jorge Contreras who lays out an alternative approach focused on the reform of standards-setting organizations (Contreras 2012a).

32. See discussion below of recent changes in the approach of the Department of Justice relative to patent policies of standards-development organizations like VITA in "Testing the Limits: The US Department of Justice Supports VITA's ex ante Disclosure of Essential Patents."

33. "Leading Global Standards Organizations Endorse 'Open Standard' Principles that Drive Innovation and Borderless Commerce," *Business Wire*, August 29, 2012.

34. The author is the president of the IEEE-Standard Association (IEEE-SA).

35. See also the most recent update of this circular, Office of Management and Budget, 2012

36. According to its website, DIN, the German Institute for Standardization, "develops norms and standards as a service to industry, the state and society as a whole. A registered non-profit association, DIN has been based in Berlin since 1917. DIN's primary task is to work closely with its stakeholders to develop consensus-based standards that meet market requirements. By agreement with the German Federal Government, DIN is the acknowledged national standards body that represents German interests in European and international standards organizations. Ninety percent of the standards work now carried out by DIN is international in nature."

http://www.din.de/cmd;jsessionid=AE8D5D82479D566B1F1E4873FC9AE 59A.3?level=tpl-bereich&menuid=47566&cmsareaid=47566&languageid=en, accessed April 27, 2010.

37. See also the more recent research findings in Mowery 2009

38. Accredited by ANSI as an American National Standards developer and a submitter of Industry Technical Agreements to IEC, VSO provides its members with the ability to develop and promote open technology standards. Standards development takes place in working groups and study groups. VITA has created more than 30 standards in the past 10 years that promote open technology systems. http://www.vita.com/vso-stds.html.

39. ANSI 2008. See also the original decision by ANSI's Executive Council, dated October 1, 2007, stating: "Motorola's Arguments that VITA's Disclosure Obligation Impermissibly Removes a RAND [reasonable and non-discriminatory] Option Guaranteed by the ANSI Patent Policy and Impermissibly Imposes a *De Facto* Duty to Conduct a Patent Search in Violation of the ANSI Patent Policy Are Without Merit," http://www.vita.com/disclosure/ANSI%20ExSC%20Panel% 20Decision%20in%20Motorola%20Appeal%20of%20VITA%20 Reaccreditation%20001Oct07.pdf, accessed May 27, 2010.

40. This section has greatly benefited from comments provided by Jorge L. Contreras and his path-breaking analysis in Contreras 2012b.

41. For a economic analysis of the impact of these new chip design methodologies see Ernst 2005 a and 2005 b.

42. Author's interviews and "The Networked Grid 100: Movers and Shakers of the Smart Grid," *Greentech Media*, February 1, 2010. http://www.greentechmedia.com/articles/read/the-networked-grid-100.

43. The American Recovery and Reinvestment Act of 2009 earmarked $4.3 billion for the Smart Grid, most of it for demonstration projects and existing deployments—public-private matching funds are expected to bring total funding to $8.6 billion.

44. "2012 Smart Grid Year in Review", *Greentech Media*, December 19, 2012, http://www.greentechmedia.com/articles/read/2012-smart-grid-year-in-review

Bibliography

Abbate, J. 1999. *Inventing the Internet.* Cambridge, MA: MIT Press.

Agnew, P.G. 1926. "A Step Toward Industrial Self-Government," *The New Republic,* March 17, 95.

Alderman, R. 2009. "Market Inefficiencies, Open Standards, and Patents." Unpublished manuscript (VITA).

Alic, J. 2009. *Energy Innovation from the Bottom Up: Project Background Paper.* Paper prepared for the joint project of the Consortium for Science, Policy, and Outcomes (CSPO), Arizona State University, and the Clean Air Task Force (CATF), March.

ANSI. 2008. Email to Mr. Miguel Pellon, Vice President, Technology—Standards Corporate, Motorola, January 22. http://www.vita.com/disclosure/ ANSI%20Appeals%20Board%20Decision%20in%20Motorola%20 Appeal%2022Jan08.pdf, accessed May 27, 2010.

———. 2009. "ANSI Response to National Survey Questions on U.S. Standards Policies," NPC 016-2009, May 27. http://publicaa.ansi.org/sites/apdl /Documents/Standards%20Activities/Critical%20Issues/Survey-US%20 Standards%20Policies/ANSI-response-05-27-09.pdf.

———. 2010a. "ANSI Essential Requirements: Due Process Requirements for American National Standards." http://publicaa.ansi.org/sites/apdl/Documents/Standards %20Activities/American%20National%20Standards/Procedures,%20 Guides,%20and%20Forms/2010%20ANSI%20Essential%20 Requirements%20and%20Related/2010%20ANSI%20Essential%20 Requirements.pdf.

———. 2010b. "A Historical Overview: 1918–2008." http://publicaa.ansi.org/sites /apdl/Documents/News%20and%20Publications/Links%20Within%20 Stories/ANSI%20-%20A%20Historical%20Overview.pdf, accessed May 1.

———. 2010c. "United States Standards Strategy." http://publicaa.ansi.org/sites /apdl/Documents/Standards%20Activities/NSSC/USSS_Third_edition /USSS%202010-sm.pdf.

———. 2012. "Introduction to ANSI." http://www.ansi.org/about_ansi/introduction /introduction.aspx?menuid=1, accessed December 19.

Antonelli, C., ed. 2011. "The Systemic Dynamic of Technological Change: An Introductory Frame." In *Handbook on the System Dynamics of Technological Change*. Cheltenhamn, UK: Edward Elgar Publishing Ltd.

Arnold, G.W. 2011. Testimony before the Subcommittee on Technology and Innovation; Committee on Science, Space, and Technology; US House of Representatives Hearing on Empowering Consumers and Promoting Innovation through the Smart Grid, Washington, DC, September 8. http:// science.house.gov/sites/republicans.science.house.gov/files/documents/ hearings/090811_%20Arnold.pdf.

ASME. 2010. "A Brief History of ASME." http://www.asme.org/Communities /History/ASMEHistory/Brief_History.cfm

Axelrod, R., and M.D. Cohen. 1999. *Harnessing Complexity: Organizational Implications of a Scientific Frontier.* New York: The Free Press.

Baumol, W.J., and A.S. Blinder. 1991. *Economics: Principles and Policy.* 5th ed. Orlando, FL: Harcourt Brace Jovanovich.

Bradner, S. 1996. "The Internet Standards Process: Best Current Practice. Revision 3." Harvard University. http://tools.ietf.org/html/bcp.

Branscomb, L. and B. Kahin. (1995). "Standards processes and objectives for the National Information Infrastructure," *Information Infrastructure and Policy,* 4(2), p. 87–106.

Brown, A. 2010. "Microsoft Fails the Standards Test." http://www.adjb.net/post /Microsoft-Fails-the-Standards-Test.aspx, accessed March 31.

Cargill, C.F. 1989. *Information Technology Standardization: Theory, Process, and Organizations.* Bedford, MA: Digital Press.

Chandler, A.D. 1977. *The Visible Hand: The Managerial Revolution in American Business.* Cambridge, MA: The Belkap Press of Harvard University Press.

Cochrane, R.C. 1966. *Measures for Progress: A History of the National Bureau of Standards.* Washington, DC: US Department of Commerce.

Congressional Research Service. 2007. *Energy Independence and Security Act of 2007: A Summary of Major Provisions.* Prepared by Fred Sissine, Coordinator, Specialist in Energy Policy Resources, Science, and Industry Division, December 21. http://energy.senate.gov/public/_files/RL342941.pdf, accessed May 27, 2010.

Contreras, J.L. 2011. "An Empirical Study of the Effects of Ex Ante Licensing Disclosure Policies on the Development of Voluntary Technical Standards." Study prepared for the National Institute of Standards and Technology (NIST contract No. SB 134110SE1033), June 27.

———. 2012a. "Rethinking RAND: DSO-Based Approaches to Patent Licensing Commitments." Paper presented at the ITU Patent Roundtable, International Telecommunications Union, Geneva, October 10.

————. 2012b. "Standards, Patents, and the National Grid." *Pace Law Review* 32 (3): 641–75.

DeNardis, L. 2009. *Protocol Politics.* Cambridge, MA: MIT Press.

Department of Energy. 2003. "Grid 2003: A National Vision for Electricity's Second 100 Years," iii. US Department of Energy, Office of Electric Transmission and Distribution. http://www.climatevision.gov/sectors/electricpower /pdfs/electric_vision.pdf, accessed May 11, 2010.

Department of Justice. 2006. "Response to VMEbus International Trade Association (VITA)'s Request for Business Review Procedure." Letter from Thomas O. Barnett, Assistant Attorney General, US Department of Justice, to R.A. Skitol, Drinker, Biddle & Reath, LLP, Washington, DC, October 30. http://www.justice.gov/atr/public/busreview/219380.pdf.

Derthick, M., and P.J. Quirk. 1985. *The Politics of Deregulation.* Washington, DC: Brookings Institution.

Dertouzos, M.L., R.K. Lester, and R.M. Solow. 1989. *Made in America: Regaining the Productive Edge.* Cambridge, MA: The MIT Press.

Energy Independence and Security Act (EISA) of 2007. Public Law No: 110-140.

Ernst, D. 2002. "Electronics Industry." In *The International Encyclopedia of Business and Management (IEBM), Handbook of Economics,* edited by William Lazonick, 319-339. London: International Thomson Business Press.

————. 2005a. "Complexity and Internationalization of Innovation: Why is Chip Design Moving to Asia?" *International Journal of Innovation Management* 9 (1): 47–73.

————. 2005b. "Limits to Modularity: Reflections on Recent Developments in Chip Design." *Industry and Innovation* 12 (3): 303–35.

————. 2005c. "The New Mobility of Knowledge: Digital Information Systems and Global Flagship Networks." In *Digital Formations: IT and New Architectures in the Global Realm,* edited by R. Latham and S. Sassen. Princeton, NJ: Princeton University Press.

————. 2011. *Indigenous Innovation and Globalization: The Challenge for China's Standardization Strategy.* La Jolla, CA: UC Institute on Global Conflict and Cooperation and Honolulu: East-West Center [published in Chinese, Beijing: University of International Business and Economics Press].

————. 2012. "High Road or Race to the Bottom? Reflections on America's Manufacturing Futures." Unpublished manuscript. Last modified 2012.

Ernst, D., and J. Ravenhill. 2000. "Convergence and Diversity: How Globalization Reshapes Asian Production Networks." In *International Production Networks in Asia: Rivalry or Riches?,* edited by M. Borrus, D. Ernst, and S. Haggard, 226–56. London: Routledge.

Ferguson, E.S. 1974. "A Sense of the Past: Historical Publications of the American Society of Mechanical Engineers." http://www.asme.org/Communities /History/ASMEHistory/Sense_Past_Historical.cfm, accessed April 22, 2010.

Fisher, R., and L. Nirenberg. 2010. "Who is the Smart Grid Technology Leader?" http://www.rocketcap.com/wp-content/uploads/2010/05/smart-grid-IP-v3_2_RocketCap.pdf, accessed May.

Garcia, D.L. 1992. "Standard Setting in the United States: Public and Private Sector Roles." *Journal of the American Society for Information Science* 43 (8): 531–37.

———. 1993. "A New Role for Government in Standard Setting?" *Standard View* 1 (2): 2–10.

Garcia, D.L., B.L. Leickly, and S. Wiley. 2005. "Public and Private Interests in Standard Setting: Conflict or Convergence?" In *The Standards Edge: Future Generations,* edited by Sherrie Bolin, 126–30. Ann Arbor, MI: The Bolin Group.

Gawer, A., and M.A. Cusumano. 2002. *Platform Leadership: How Intel, Microsoft and Cisco Drive Industry Innovation.* Boston, MA: Harvard Business School Press.

Gawer, A., and R. Henderson. 2007. *Platform Owner Entry and Innovation in Complementary Markets: Evidence from Intel.* NBER Working Paper, National Bureau of Economic Research. http://www.nber.org/papers/w11852.pdf

Grewal, D.S. 2008. *Network Power: The Social Dynamics of Globalization.* New Haven, CT: Yale University Press.

Grove, A.S. 1996. *Only the Paranoid Survive: How to Exploit the Crisis Points that Challenge Every Company and Career.* New York and London: Harper Collins Business.

Hall, P.A., and D. Soskice, eds. 2001. *Varieties of Capitalism: The Institutional Foundations of Comparative Advantage.* London: Oxford University Press.

Hart, D.H. 1998. *Forged Consensus: Science, Technology, and Economic Policy in the United States, 1921–1953.* Princeton, NJ: Princeton University Press.

Hoffman, P. 2009. "The Tao of IETF: A Novice's Guide to the Internet Engineering Task Force." http://tools.ietf.org/rfcmarkup?doc=fyi17.

Hunt, R.M., S. Simojoki, and T. Takalo. 2007. "Intellectual Property Rights and Standard Setting in Financial Services: The Case of the Single European Payments Area," Working Paper no. 07-20, Research Department, Federal Reserve Bank of Philadelphia, 3. http://www.phil.frb.org/research-and-data/publications/working-papers/2007/wp07-20.pdf

Hurwitz, M. 2004. "United States Standardization Strategies and Their Relationship to ISO's Long-Term Strategy," paper presented at the ISO International Standardization Forum, Tokyo, October 25, 2004.

Institute for Defense Analyses. 2012. "Emerging Global Trends in Advanced Manufacturing," report prepared for the Office of the Director of National Intelligence (ODNI), IDA Paper P-4603, Alexandria, Virginia.

International Trade Administration. 2009. "The Voluntary Standards System: A Dynamic Tool for U.S. Economic Growth and Innovation," US Department of Commerce seminar program, July 24, 2009. Washington, DC.

Katz, M. and C. Shapiro. 1985. "Network Externalities, Competition and Compatibility." *American Economic Review* 75 (3): 424–40.

Kindleberger, C. 1983. "Standards as Public, Collective and Private Goods," *Kyklos*, Vol.36, issue 3: pages 377-96.

Kirkland, E.C. 1961. *Industry Comes of Age: Business, Labor, and Public Policy, 1860–1897.* New York: Holt, Rinehart and Winston.

Kogut, B., and U. Zander. 1993. "Knowledge of the Firms and the Evolutionary Theory of the Multinational Corporation." *Journal of International Business Studies* 24 (4).

Lemley, M.A. 2002. "Intellectual Property Rights and Standard-Setting Organizations." *California Law Review* 90: 1889–981.

———. 2007. "Ten Things to Do About Patent Hold-Up of Standards (and One Not to)." *Boston College Law Review* 48:149–68.

Lemley, M.A., and C. Shapiro. 2007. "Patent Holdup and Royalty Stacking." *Texas Law Review* 85:1991–2041.

Libicki, M. 1995. *Information Technology Standards: Quest for the Common Byte.* Rockport, MA: Digital Press.

Libicki, M., J. Schneider, D.R. Frelinger, and A. Slomovic, 2000. *Scaffolding the Web: Standards and Standards Policy for the Digital Economy.* Santa Monica, CA: RAND Science and Technology Policy Institute.

Lord, P.E. 2007. "Risky Business: The US Software Industry's Perspective on US Government Engagement in the Process of Standard Setting." MA thesis, Graduate School of Arts and Sciences, Georgetown University.

Mattli, W., and T. Buethe. 2003. "Setting International Standards: Technological Rationality or Primacy of Power?" *World Politics* 56:1–42.

Mills, S. 2012. "International Standards in the Emerging Global Economy: IEEE-SA." http://open-stand.org/wp-content/uploads/2012/11/International-Standards-in-the-Emerging-Global-Economy-V2.pdf.

Mowery, D.C. 2009. "Plus Ça Change: Industrial R&D in the "Third Industrial Revolution." *Industrial and Corporate Change* 18 (1): 1–50.

Mowery, D.C., and R.R. Nelson, eds. 1999. *Sources of Industrial Leadership.* Cambridge, UK: Cambridge University Press.

National Science Board. 2012. *Science and Engineering Indicators 2012.* Arlington, VA: National Science Foundation.

National Science and Technology Council. 2011. "A Policy Framework for the 21st Century Grid: Enabling Our Energy Future." http://www.whitehouse.gov/sites/default/files/microsites/ostp/nstc-smart-grid-june2011.pdf.

NIST. 2010. *Framework and Roadmap for Smart Grid Interoperability Standards, Release 1.0.* Office of the National Coordinator for Smart Grid Interoperability, NIST Special Publication 1108. Washington, DC: National Institute of Standards and Technology, US Department of Commerce.

———. 2012. *NIST Framework and Roadmap for Smart Grid Interoperability Standards, Release 2.0*. NIST Special Publication 1108R2. Washington, DC: National Institute of Standards and Technology, US Department of Commerce.

Office of Management and Budget. 1998. *Circular A-119*. Washington, DC: Office of Management and Budget.

———. 2012. *Memorandum for the Heads of Executive Departments and Agencies*, January 17. Washington, DC: Office of Management and Budget.

Office of Technology Assessment. 1992. "Standards Setting in the United States." In *Global Standards: Building Blocks for the Future*. Washington, DC: Office of Technology Assessment.

Olson, M. 1971. *The Logic of Collective Action: Public Goods and the Theory of Groups*. Revised ed. Cambridge, MA: Harvard University Press.

Ostrom, E. 1990. *Governing the Commons: The Evolution of Institutions for Collective Action*. Cambridge, UK: Cambridge University Press.

Powers, C. 2009. "Public-Private Interaction in the US Standards System: New Challenges in the Global Knowledge Economy." In *Standards and Innovation Policy in the Global Knowledge Economy: Core Issues for China and the US*. Proceedings of the Beijing Conference, 8–11. Honolulu: East-West Center and Washington, DC: National Bureau of Asian Research.

Rohlfs, J.H. 2001. *Bandwagon Effects in High-Tech Industries*. Cambridge, MA: MIT Press.

Russell, A.L. 2005. "Standardization in History: A Review Essay with an Eye to the Future." In *The Standards Edge: Future Generations,* edited by Sherrie Bolin, 247–60. Ann Arbor, MI: The Bolin Group.

———. 2006. "Industrial Legislatures: The American System of Standardization." In *International Standardization as a Strategic Tool*, 70–79. Geneva: International Electrotechnical Commission (IEC).

SAC. 2004. *Study on the Construction of National Technology Standards System*. Beijing: Standardization Administration of China.

Scherer, F.M. 1977. *The Economic Effects of Compulsory Patent Licensing*. New York: New York University Graduate School of Business Administration.

———. 2006. "The Political Economy of Patent Policy Reform in the United States." http://www.nber.org/~confer/2007/si2007/PRL/scherer.pdf.

Sheremata, W.A. 2004. "Competing Through Innovation in Network Markets: Strategies for Challengers." *Academy of Management Review* 29 (3): 359–77.

Simcoe, T. 2007. "Delay and de jure Standardization: Exploring the Slowdown in Internet Standards Development." In *Standards and Public Policy*, edited by S. Greenstein and V. Stango, 260–95. Cambridge, UK: Cambridge University Press.

———. 2012. "Standard Setting Committees: Consensus Governance for Shared Technology Platforms." *American Economic Review* 102 (1): 305–36.

Sinclair, B. 1980. *A Centennial History of the American Society of Mechanical Engineers: 1880–1980.* Toronto: Toronto University Press.

Smith, A. 1776. *An Inquiry Into the Nature and Causes of the Wealth of Nations,* book 1, chapter III. Reprinted 1970. Harmondsworth, Middlesex,UK: Penguin Books.

Spring, M.B. 2009. "What Have We Learned about Standards and Standardization?" Unpublished manuscript, School of Information Sciences, University of Pittsburgh.

St. John, J. 2012. "2012 Smart Grid Year in Review." *Greentech Media.* http://www.greentechmedia.com/articles/read/2012-smart-grid-year-in-review, December 19.

Stango, V. 2004. "The Economics of Standards Wars." *Review of Network Economics* 3 (1): 1–19.

State Grid Corporation of China. 2011. "Smart Grid Implementation & Standardization in China." http://www.smartgrid.com/wp-content/uploads/2011/09/6___Changyi.pdf, November.

Steinfield, C.W., R.T. Wigand, M.L. Markus, and G. Minton, 2007. "Promoting E-Business Through Vertical IS Standards: Lessons From the US Home Mortgage Industry." In *Standards and Public Policy,* edited by S. Greenstein and V. Stango. Cambridge, UK: Cambridge University Press.

Tapia, C.G. 2010. *Intellectual Property Rights, Technical Standards and Licensing Practices (FRAND) in the Telecommunications Industry.* Koeln: Carl Heymans Verlag.

Tassey, G. 2007. *The Technology Imperative.* Cheltenham, Gloucestershire: Edward Elgar Publishing Ltd.

Tate, J. 2001. "National Varieties of Standardization." In *Varieties of Capitalism: The Institutional Foundations of Comparative Advantage,* edited by P.A. Hall and D. Soskice. Oxford, UK: Oxford University Press.

Updegrove, A. 2008. "Behind the Curve: Addressing the Policy Dependencies of a 'Bottom-Up' Standards Infrastructure." *Standards Today,* October–November: 1–26.

———. 2009. "How We'll Get the Job Done: An Interview with NIST's Dr. George W. Arnold." *Standards Today* VIII (3).

———. 2012. "Openness and Legitimacy in Standards Development." *Standards Today* XI (1): 1–6.

Weiss, M.B.H., and M.B. Spring. 2000. "Selected Intellectual Property Issues in Standardization." In *Information technology Standards and Standardization: A Global Perspective,* edited by Kai Jacobs, 63–79. Hershey, PA, and London: Idea Group Publishing.

Wolff, A.W. 2010."The Direction of China's Trade and Industrial Policies." Testimony before the House Ways and Means Committee, US House of Representatives, Washington, DC, June 16.

Yamauchi, Toru. 2004. "Comprehensive Strategy for International Standardization Activities for Japan." http://www.cicc.or.jp/japanese/hyoujyunka/pdf_ppt/04SEJapanese%20Standardization%20Polocy.pdf, accessed April 30, 2010.

Ziegler, K. 2011. "Chinese Standardization in Smart Grids: A European Perspective." *Talk Standards.* http://www.talkstandards.com/chinese-standardization-in-smart-grids-a-european-perspective/, May 10.

Acknowledgments

I have benefited from comments, ideas, and suggestions from a diverse group of analysts, business executives, journalists, policymakers, and scholars in Asia, Europe, and the United States.

For introducing me into the complex world of standardization, I owe a particular debt of gratitude to Jorge L. Contreras, Linda Garcia, Klaus Ziegler, Carl Cargill, Chuck Powers, Michael B. Spring, Wang Ping, Song Mingshun, Michael Ding, Andrew Updegrove, Gregory Tassey, Greg Shea, Scott Kennedy, Pete Suttmeier, and Nathaniel Ahrens. On the interface between intellectual property rights and standardization, I have benefited greatly from discussions with and the writings of Richard R. Nelson, David C. Mowery, Joseph Straus, Tim Simcoe, Konstantinos Karachalios, An Baisheng, Claudia Tapia, Ray Alderman, Cliff Reader, Thomas Pattloch, Heinz Goddar, Georg Kreuz, Alan Fan Zhiyong, Michael Ding, Wang YiYi, and Zhang Yan. For insights into the dynamics of innovation policymaking in the United States, I am grateful in particular to Willy Shih, Susan Helper, Stephanie S. Shipp, David M. Hart, Chris Hill, John Alic, Paul A. Lengermann, David M. Byrne, Rob Atkinson, Stephen Ezell, Kent Hughes, Michael Lind, Jimmy Goodrich, David Hoffman, Bart van Ark, and James Fallows.

I am very fortunate to work at the East-West Center and I thank Charles E. Morrison and Nancy Lewis for supporting this research. Thanks also to East-West Center colleagues, especially Toufiq Sidiqqi and Richard Baker, for brainstorming discussions. This study has greatly benefited from discussions and seminars organized as part of the China Innovation MINERVA project at the Institute on Global Conflict and Cooperation, University of California at San Diego. I am particularly grateful to Susan Shirk, Barry Naughton, TaiMin Cheung, and Heidi Serochi. In addition, I have greatly benefited from discussions and seminars organized by the Board of Global Science and Technology of the National Academies; the Board on Science, Technology, and Economic Policy of the National Academies; the Science and Technology Policy Institute at the Institute of Defense Analyses (Washington, DC); the Innovation and Information Technology Foundation (Washington, DC); CENTRA Technology (Arlington, VA); the Center for Science and Technology Policy at George Mason University; the School of Public Policy and Management, Tsinghua University (Beijing); the China National Institute for Standardization (Beijing); the Taiwan Institute of Economic Research (Taipei); the Chunghua Institution for Economic Research (Taipei); and the Industrial Technology Research Institute (Hsinchuh, Taiwan).